AMERICAN REPRESENTATION IN OCCUPIED GERMANY,

1922 - 1923

COMPILED BY THE ASSISTANT CHIEF OF STAFF, G-2

AMERICAN FORCES IN GERMANY

AMERICAN REPRESENTATION IN OCCUPIED GERMANY

1922 - 1923

P R E F A C E

The present work, contained in one volume, is the continuation of the history prepared by Major Philip H. Bagby, Infantry, under the title of "American Representation in Occupied Germany, 1920-21". It also represents the last volume of the account of the American participation in the occupation of the Rhineland. The period covered extends from January 1st, 1922, to February 3rd, 1923.

In order to obtain a comprehensive picture of the American representation in the Rhineland, it will be necessary to peruse, in addition to this volume and the two volumes of 1920-1921, Colonel Irvin L. Hunt's four-volume treatise entitled "American Military Government of Occupied Germany, 1918-1920". Policies and functions of organizations and activities described fully in either of the preceding works are not covered again in detail, but as a rule referred to by foot-notes, so that an exact understanding of the part played by America in the Rhineland may be obtained only if the previous six volumes be available.

The arrangement devised for the 1920-1921 history has in general been retained. Chapter I serves to familiarize the reader with the political and economic situation in unoccupied Germany and the occupied zone, without which this account could not be considered complete. This deals not only with the general political and economic history of 1922, the "year of reparations notes", but also with all events leading up to the situation existing in the Rhineland on the date of issue of this work. In addition, all items of interest in regard to the movements of the Commanding General, American Forces in Germany, and the evacuation of the American troops from the Rhineland are noted therein. Chapter II deals with the activities and outstanding policies of the Interallied Rhineland High Commission. Chapter III describes the other agencies whereby the relations of the Army with the civil population of the American area were maintained and regulated. The last part of the work - Chapter IV - concerns itself with those activities of the American Forces in Germany concerning the military only, and depicts the gradual reduction of our military strength on the Rhine until only a very small force was left, following the departure of the 8th Infantry. Map No. 1, showing the American zone as it existed on the date of withdrawal, and map No. 2, showing the Ruhr District, are inserted at the back of the volume.

This history was written by, and compiled under the supervision of, Lieutenant Colonel Henry Hossfeld, Assistant Chief of Staff, G-2, of the American Forces in Germany, who was assisted by Captain Bernard B. McMahon, Infantry, Captain Joe N. Dalton, Infantry, 1st Lieutenant Clarence Bird, M.I.D., O.R.C., and Dr. A. P. Happel, I.A.R.H.C.

On January 11th, 1923 - the date of receipt of the withdrawal order - instructions were issued to all sections and services of these Headquarters to furnish an account of their activities for the period covered by the history. Upon receipt of these reports the actual writing was undertaken. The time of compilation and writing of the history was one month to a day, and the first copy was handed to the Commanding General on the date of its issue - February 19th - just prior to his departure for the United States.

Coblenz, Germany,

February, 1923.

Abbreviations.

A. C. of S. .... Assistant Chief of Staff.

A. D. C. .... Aide-de-Camp.

A. F. G. .... American Forces in Germany.

A.M.G. of O. G. .... American Military Government of Occupied Germany.

A. R. in O. G. .... American Representation in Occupied Germany.

Centrum .... Centrum Party.

C. L. of C. .... Commander, Line of Communication.

Comm. .... Communist Party.

C. T. O. .... Chief Transportation Officer.

D. C. I. .... Division of Criminal Investigation.

Dem. .... German Democratic Party.

D. G. C. R. A. .... Direction Generale de Communications et de Ravitaillement d'Armee (Hq. Army Communications and Supplies).

F. & R. O. .... Fiscal and Records Office.

G-1, G-2, G-3, G-4 .... Divisions of General Staff:
G-1 - Administration and Personnel.
G-2 - Military Intelligence.
G-3 - Operations and Training.
G-4 - Supply.

G. N. P. P. .... German National People's Party.

G. P. P. .... German People's Party.

I. A. R. C. .... Interallied Railways Commission.

I. A. R. H. C. .... Interallied Rhineland High Commission.

I. A. W. C. .... Interallied Waterways Commission.

L. of C. .... Line of Communication.

O. C. C. A. .... Officer in Charge of Civil Affairs.

O. D. T. .... Ordre de Transport.

P. M. D.	....	Provost Marshal's Department.
R. R. and C.	....	Rents, Requisitions and Claims.
R. T. O.	....	Railway Transportation Officer.
U. S. P.	....	United Socialist Party.

TABLE OF CONTENTS.

AMERICAN REPRESENTATION IN OCCUPIED GERMANY.

1922 - 1923.

CHAPTER I.

R E V I E W O F E V E N T S.

——— ooOoo ———

CHAPTER II

ACTIVITIES OF INTERALLIED RHINELAND HIGH COMMISSION.

CHAPTER III

Page

(Photographs and Maps ommitted)

TABLE OF CHARTS.

PHOTOGRAPHS and MAPS.

-----ooOoo-----

CHAPTER I

R E V I E W O F E V E N T S

Page

------------ ooo 0 ooo ------------

REVIEW OF EVENTS.

First Quarter of 1922.

Preparations for Cannes Conference.

The political and economic situation in Germany at the beginning of 1922 was still dominated by the reparations obligations under the Versailles treaty and subsequent agreements. The Reparations Commission had replied unfavorably to Chancellor Wirth's notice that the government would be unable to render the next payments in full and that a moratorium was absolutely necessary on the balance. The British and French Premiers had held conversations at London and decided to call a conference at Cannes early in January, when German reparations were to be discussed in connection with economic reconstruction and reestablishment of normal conditions. In the meantime, experts were making preparatory arrangements at London and Paris.

The international difficulties which the Wirth government faced after it had proclaimed Germany's inability to abide by the London agreement failed to bring about a cabinet crisis. Negotiations seeking the extension of the national government, so as to include the German People's Party, were deadlocked. The government was at the same time making every effort to bring the parties of the Right and Left together in order to have them agree upon some taxation compromise, since the taxation issue was the internal problem of the day as 1922 began.

Taxation Reform.

The German cabinet took up pressing domestic questions, especially the taxation problem, with leading representatives of all the political parties early in January. The Reichstag Committee began its deliberations on January 10, in order to be able to submit the taxation bill to the plenary session of the House, which was to convene on January 19. The solution of the taxation problem soon became the most important issue. The Chancellor made every effort to reconcile the divergent views of the political parties, but only after the cabinet had threatened to resign did he succeed in bringing about a compromise between the government coalition parties and the German People's Party.

Chancellor Wirth's Policy of Fulfilment.

The principal points of the taxation compromise were outlined by Chancellor Wirth before the Reichstag on January 26. The most important feature of the proposed bill was the floating of a compulsory loan amounting to 1,000,000,000 gold marks. The Chancellor declared that the policy of fulfilment was the only one which could succeed. He asserted, however, that the German government would never accede to the demand for the surrender to the Allies of alleged war criminals. He concluded by advocating participation in the reconstruction of Russia.

It was generally conceded that the Chancellor had gained in prestige as a result of his speech. Nearly all the parties supported the compromise. Even the German Nationalists refused to join in the vote of lack of confidence in Dr. Wirth which was introduced by the Communists but overwhelmingly defeated

Party Affairs.

The Independent Socialist convention which opened at Leipsic on January 8 debated the taxation question, the class struggle and unification of effort. The convention opposed any union with the Bourgeois parties and advocated dictatorship of the proletariat, declaring that a final consolidation of the working-classes would take place as soon as the Majority Socialists were ready to consider the interests of the proletariat only.

The National Centrum Party held its annual convention at Berlin from January 15th to 17th. A new platform was unanimously adopted which emphasised the Christian character of the party and strongly advocated unity of the German states. Dr. Wirth addressed the convention and was given a vote of confidence.

Other conventions were held in January by the Rhenish Democrats and the German Anarchists. The Rhenish Separatists showed renewed vigor in their agitation for an independent Rhineland. An anti-Separatist demonstration was held at Berlin by Rhinelanders residing there.

The Communist Party lost ground. Several Reichstag

deputies left the ranks of the party and joined the Levi group of the Communist Workers' Union, which then controlled fifteen seats in the Reichstag, whereas the Communists retained but twelve.

To judge from the sale of one of its newspapers to the German People's Party, the influence of the Democratic Party also seemed to be waning at this time. The German People's Party was considering the advisability of entering the coalition government, which the Democrats rejoined when Dr. Rathenau was appointed Minister of Foreign Affairs on January 31.

Reparations.

Early in January the Reparations Commission explained to a German delegation, headed by Dr. Fischer, that the Commission could not consider the German request for a moratorium until the German reply to the Commission's note of December 16 had been received. The entire question was therefore postponed pending the result of the Cannes conference.

Cannes Conference.

The importance of the German reparations payments in connection with the general situation in Europe having been recognised by the Supreme Council, the German government was invited to send delegates to an economic conference at Cannes on January 6. Germany accepted the invitation, and Dr. Rathenau was chosen to head its delegation.

In speaking before the Supreme Council and the Reparations Commission, Dr. Rathenau again emphasised Germany's inability to make the reparations payments which were to fall due on January 15 and February 15. The guarantees he offered, and which comprised increased coal prices, higher postal and telegraph rates and the suspension of certain allowances, were regarded by the Allies as somewhat below what they had expected.

The resignation of M. Briand, French Premier, brought the conference to an abrupt end. No final decision was reached by the Supreme Council regarding the reparations question; but the reply of the Reparations Commission to the German request for a partial moratorium on the payments due on January 15 and February 15 was announced as a temporary measure. A preliminary respite was granted on condition that the German government make a payment of 31,000,000 gold marks every ten days, that it submit an adequate plan of financial reform within a fortnight and simultaneously propose a program of reparations payments covering the current year. This temporary arrangement was to cease as soon as the Reparations Commission had finally passed on the German proposals.

German Note to Reparations Commission.

On January 28, Chancellor Wirth dispatched a note to the Reparations Commission in which he outlined the contemplated reforms of the financial administration and proposed a new plan for the reparations payments. The note declared that in an effort to augment receipts, the "business turnover", sugar and coal taxes would be increased and customs collected on a gold basis. In order to guarantee the 1922 payments, a compulsory loan of 1,000,000,000 gold marks, bearing interest only at the end of three years, would be resorted to. Expenditures were to be curtailed by a cut in the number of government employees, the non-payment of subsidies on food supplies and by a balancing of the deficits of the railway and postal administrations. A bill would be introduced to make the Reichsbank independent of the national government, and a voluntary loan would be floated to prevent further increase of the floating debt. The note further declared that deliveries in kind were preferable to payments in specie, and that the German government was desirous of concluding agreements with other nations similar to the agreement with France entered into at Wiesbaden.

Leipsic Verdicts Rejected.

In a resolution submitted to the Supreme Council, the Interallied Commission, which had been investigating the Leipsic trials of war criminals, rejected the findings and sentences of the German Supreme Court and recommended the extradition of the alleged war criminals to the Allied powers. Chancellor Wirth declared in a speech on January 26 that such a demand would never be complied with.

Railway Strike.

The negotiations regarding the wage increases demanded by the railway employees having been carried on throughout January without the desired result, the National Federation of Railway Employees proclaimed a strike on February 2. This affected the entire country except Upper Silesia, the Rhineland and parts of Southern Germany. The strike lasted about one week, when a compromise was effected between the government and the employees. The government refused to deal with any striking organisation but agreed to continue negotiations for the purpose of readjusting the pay scale and to refrain from punishing those who had participated in the strike, except in the case of the leaders.

The sympathy strike of the Berlin municipal employees was settled on February 8, after having caused great hardship, the capital having been without water, gas, heat and electricity for several days.

Clash in Reichstag.

The strong support given the strikers by the two radical parties of the left turned the strike issue into a test of strength between the government and radicalism. In defining the government's attitude, the Chancellor declared that the railwaymen, being government employees with pension privileges, had no right to strike. The stormy Reichstag debate following this statement resulted in four separate motions for a vote of lack of confidence in the government. These were introduced by the Communists, the Independent Socialists, the German People's Party and the German National People's Party. In order to decide the issue, the coalition parties, at the Chancellor's request, introduced a motion for a vote of confidence on February 15. Due to indirect support by the Bavarian People's Party, which largely refrained from voting, and a split in the Independent Socialist ranks, Dr. Wirth received a scant majority of 35 votes. For the time being the crisis was passed, and the dissolution of the Reichstag, which was apparently contemplated in case of Dr. Wirth's defeat, was prevented.

Boulogne - Genoa.

The German government received an official invitation in January to attend an international economic conference scheduled to be held at Genoa, Italy. The invitation was accepted and hopes ran high for a time. However, faith in the success of the conference began to ebb in February, when a postponement was announced and it was becoming more and more apparent that the United States would not participate in it. In addition, the preliminary conversations between M. Poincare and Mr. Lloyd George, French and British Prime Ministers, held at Boulogne, did much to discourage the Germans and were unfavorably commented upon by the German press, which saw in the agreement reached there between France and England a limitation of the scope of the conference, especially at Germany's expense. These two distinguished statesmen met on February 25 and agreed upon a program for the conference which may be summarised as follows:

1. The Genoa conference will not endanger the Versailles treaty or any other agreement concluded by France and the Allies;

2. France's rights under the treaty, especially in regard to the adoption of coercive measures, will not be disputed;

3. The conference will in no way abolish the claims of the Allies;

4. The opening of the conference will be postponed until April 10, 1922;

5. The rights and privileges of the League of Nations will not be curtailed.

Vistula Award.

The Interallied Delimitation Commission's decision to award to Poland five West Prussian communities on the east bank of the Vistula River created great indignation throughout Germany. It was contended that this ruling was contrary to the terms of the Versailles treaty. The Reichstag passed a resolution protesting against it and a public demonstration was held at Berlin.

New Appointments.

Several new appointments to important public offices were made in March. Dr. Hermes, Minister of Food and Agriculture, was appointed Minister of Finance on March 10. On March 31, Reichstag Deputy Anton Fehr, of the Bavarian Peasant's League, was appointed to the post left vacant by Dr. Hermes. Herr Otto Wiedfeldt, Director of the Krupp Works, was chosen ambassador to the United States.

Disarmament Notes.

The Ambassadors of the Allied governments at Berlin forwarded a joint note to the German government on March 15. In it, they complained of Germany's failure to comply with its disarmament obligations and condemned the recalcitrant attitude of the German government towards certain instructions issued by the Interallied Military Control Commission with regard to the Reichswehr and the Schutzpolizei (Army and Security Police). The German government was requested to straighten out the differences without delay. In reply to a German explanatory note, another note concerning in Schutzpolizei was forwarded to the

Berlin government by General Nollet, Chairman of the Control Commission.

Allied Note of March 21.

After repeated intimations on the part of Germany that the ten-day payments of 31,000,000 gold marks were becoming increasingly difficult, the Reparations Commission at length made a reply to the German note of January 28, which proposed financial guarantees and a plan of payment for the current year. In this note of March 21, the Commission declared that the German proposals were inadequate, that 60,000,000,000 paper marks in addition to the money to be secured through the taxation compromise should be raised in 1922 and that the payments for 1922 should be 720,000,000 gold marks in cash and 1,450,000,000 gold marks in deliveries in kind. Severe guarantees, which practically constituted a control of Germany's financial administration, were also demanded. The note declared that the London plan of payments would be applied in case of Germany's non-compliance.

The note made a deep impression on the German public, which generally considered the terms impossible of fulfilment. Towards the end of the month, however, a more optimistic opinion prevailed, and it was believed that they might form the subject of further negotiations.

Chancellor's Reply in Reichstag.

Chancellor Wirth refused the Allied demands in a speech before the Reichstag on March 28. He was warmly supported. A vote of confidence introduced by the coalition parties was passed by an overwhelming majority, the G.P.P. and the Bavarian People's Party falling in line and the Independent Socialists refraining from voting. The Chancellor's position was greatly strengthened.

Cost of Living.

The economic situation in Germany during the first part of 1922 grew worse constantly. The official bread prices in all cities were increased in February. Potatoes were a bit scarce during the first two months of the year, but were again offered for sale in sufficient quantities in March. The average price for a kilogram was from 4 to 5 marks. Meat prices, also, made sensational advances. Beef, after an increase of from 2 to 4 marks in January and from 7 to 12 marks in February, made a jump of from 12 to 18 marks in March. Compared with May, 1921, meat prices had doubled.

The general upward trend of the cost of living during this period may be fully realised when the index figures for wholesale prices are examined, assuming that the average 1913 price equal 100. The figures follow, with those of the United States added for comparison:

	March, 1921	Jan., 1922	Feb., 1922	March, 1922
Germany	1,338	3,665	4,103	5,229
United States	123	124	126	126

Fuel.

The total production of coal during the first quarter of 1922 increased in both the Ruhr District and in Upper Silesia. The daily output, also, was greater than in December, but compared with the average of 1913, the 1922 figures still lagged far behind, in spite of the increased number of miners.

The railway strike, which lasted from February 1 to 9, and the freezing of waterways, rendered the transportation of coal almost impossible for a time. This naturally resulted in a material increase of the reserve stock on hand at the mines. In the Ruhr, this increase was from 578,600 tons at the end of January to 1,004,800 tons at the beginning of March, 1922. During the latter month, however, conditions returned to normal.

Shipping.

After the complete cessation of German shipping during the years of war and blockade, 1920, and especially 1921, had brought a vigorous recovery. The ports of Hamburg, Cuxhaven and Emden, in particular, made great strides and the 1921 tonnage of those harbors was double that of 1920. The number of ships did not grow so rapidly because of the fact that, on the whole, bigger vessels were used. Although much was regained, still pre-war figures were much higher. Hamburg's 1921 tonnage, for instance, amounted to only 65.8% of the 1913 total.

The figures for Hamburg, Antwerp and Rotterdam for January and February, 1921 and 1922, are given below for comparison:

	Hamburg		Antwerp	
	Boats	Net Registered Tons	Boats	Net Registered Tons
1922	1.165	1,591,363	1,209	1,764,386
1921	1,068	1,328,225	1,174	1,722,259
Plus:	97	263,138	35	42,127

Rotterdam

	Boats	Net Registered Tons
1922	1,106	1,559,441
1921	1,256	1,693,592
Minus:	150	134,151

Foreign Trade.

While imports decreased slowly during the first quarter of 1922, the volume of exports left much to be desired. The counter-measures which had been adopted by many countries in order to prevent flooding of their markets with German goods began to make themselves felt. However, in point of value, German exports exceeded imports both in January and February.

Deliveries in Kind.

An agreement for the payment of reparations by means of deliveries in kind was concluded between representatives of the German government and M. Benelmans, Belgian member of the Reparations Commission, in March. The Reparations Commission accepted the four agreements dealing with deliveries in kind, i.e., the Wiesbaden agreement, that regarding restitution, the Benelmans agreement and the Franco-German annex to the Wiesbaden pact, at the same time.

Depreciation of German Mark.

The period from December, 1921, to April, 1922, was marked by great fluctuations in the value of the mark. At the beginning of December, it rose from 242 to 185 to the dollar within 24 hours. Large fortunes which were involved in speculation vanished over night. A panic ensued in which many small banks and exchanges were ruined. During January, February and March the mark gradually and steadily depreciated in value as the ten-day reparations payments of 31,000,000 gold marks were made. At the end of March, upon publication of the news that the Reparations Commission considered the German proposals inadequate and demanded the imposition of additional taxes amounting to 60,000,000,000 paper marks, it took a sudden slump and reached 340 to the dollar. However, the government declared these demands impossible of fulfilment, and it improved slightly.

The Reichsbank.

In accordance with the intention expressed in its January note, the German government made an effort to reduce paper inflation, and the number of Reichsbank notes in circulation was actually decreased soon by 3,000,000,000 marks. Normal banking relations were established with the Bank of England, the Bank of the Netherlands and other important foreign banks. However, when the railway strike began in February, the floating debt increased to such a marked degree that it necessitated further printing of paper marks. From then until the end of March, the currency inflation continued. On March 31, the amount of Reichsbank paper in circulation was 130,000,000,000. The law providing for autonomy of the Reichsbank was proposed in the Reichsrat in March, but final action was postponed for some time.

Compulsory Loan.

A final compromise on the taxation legislation was

reached between the government coalition parties, the Bavarian People's Party and the German People's Party on March 9th, when they agreed to enforce a compulsory loan of 1,000,000,000 gold marks. The G.P.P. demanded that the money thus raised be used for reparations payments only. At the end of March, on the eve of the Genoa conference, the taxation bill passed its second reading and was awaiting passage.

Low Water Level of Rhine.

Much difficulty was experienced by Rhine shippers late in 1921 on account of the unusually low water level, which hardly permitted one sixth of the normal amount of fuel to be shipped. Although the railways made every effort to compensate for the adverse conditions prevailing upon the waterway, it was at times feared that the American area would experience a serious coal shortage. In January of the new year, however, the weather changed, bringing abundant rain and snow, which resulted in a very considerable rise of the Rhine and facilitated shipping greatly. With the exception of the disadvantage of heavy ice floes during February, shipping on the Rhine remained normal thereafter.

In Occupied Zone.

Separatism.

The activities of the Separatists of all shades continued, without, however, netting any appreciable gains. The most prominent and energetic of the leaders was Herr Joseph Smeets, in whose favor the interallied Rhineland High Commission intervened with the German authorities in December, ordering them to release him from the arrest in which they were holding him pending an investigation.[1] The storm of protest which followed this action by the Commission served Herr Smeets as a welcome means of advertising, and he had little trouble in remaining in the lime-light.

Smeets Trial.

His trial, on the charge of insulting German public officials, was finally scheduled to be held at Cologne on February 15th. The interallied Rhineland High Commission decided to permit it, providing that no sentence be passed without its sanction. The trial was consequently proceeded with and Herr Smeets was sentenced to three weeks' imprisonment. Later, he was again tried for having insulted the National President and, on March 3rd, sentenced again, this time to 5 months' imprisonment, the term to include the original three weeks. The sentences were not executed, pending a decision by the Rhineland High Commission.

Control of Schools by I.A.R.H.C.

A report that the Rhineland High Commission intended to a certain extent to control the nomination of candidates for positions as public school teachers in the occupied area was published in the Coblenz press under big headlines early in January. It caused much adverse comment. Professor Moldenhauer, Rhenish Reichstag Deputy of the German People's Party,

---

[1] A.R. in O.G., 1920-21, Vol. I, page 141

referred to the Commission's new policy in an inquiry submitted to the national administration, in which he declared that the Rhineland Commission had decided to form a special committee for the purpose of supervising instruction in German Schools. The question was also debated in the Prussian Assembly and continued to evoke comment in the entire German press. Later, the press published a statement said to have been made by the Interallied Rhineland High Commission, to the effect that no supervision of public instruction was intended, but merely a temporary investigation for the purpose of ascertaining whether the instruction was being imparted in accordance with the spirit of reconciliation provided for by the Weimar constitution.

In American Area.

Political Activities.

Political activities in the American area were carried on in the usual manner during this period, all parties holding meetings of more or less importance at regular intervals. The Independent Socialists evidently believed the time opportune for making an attempt to get a footing in Coblenz, where they had never been successful thitherto; but the failure of the first meeting convinced them that public opinion had not changed in their favor.

Employment.

The employment situation in the American area remained favorable all through this period, the number of unemployed being negligible. The special industries, such as the manufacture of firebricks and the clay pits of the bridgehead, had to suspend operation for a time because of a heavy snowfall. The number of unemployed was, however, hardly influenced by this enforced idleness, most of the seasonal workers finding jobs in the brown coal mines south of Cologne.

Railway Strike at Coblenz.

Coblenz found itself face to face with the possibility of a serious railway strike in the last week of 1921. Following the example set by the Cologne men, the workers of the Coblenz section of the railway system walked out and seemed determined to enforce their demands for more pay. The local branch of the German Railwaymen's Union dispatched messages to the national Ministers of Transportation and Finance as well as to the Chancellor, to the effect that the strikers in the American area were determined to hold out to the end. The messages further pointed to the serious domestic and foreign political consequences of an extended strike movement and placed the entire responsibility for the situation on the shoulders of the government. These messages, however, although signed by Herr Kalt, the union leader, were said not to represent the viewpoint of the majority of the Union's members. The Christian Trade Unions published a proclamation to the railwaymen, disapproving of the strike movement on the grounds that it was a so-called "wild strike," i.e., one entered upon without authority of the central organisations. The proclamation went on to say that Communist elements had become the real leaders of the movement. The military authorities intervened and the men were ordered to return to work. In the meantime, a conference was held by the central organisations at Berlin, as a result of which the movement was brought to an end on January 1st, 1922.

Labor Disputes.

With the exception of the railway strike at the beginning of the year, the American area remained free from serious wage disputes during the first part of this period. Such differences as arose were always quickly settled by agreement or arbitration. The big railway strike which occurred in practically all of unoccupied Germany early in February did not spread to the occupied territory. In connection with this strike, the Interallied Rhineland High Commission forwarded a note to German National Commissioner Prince v. Hatzfeldt[1] regarding maintenance of order in case of an emergency. The note called attention to the fact that in special instances the Commission was entitled to authorise the Commander-in-Chief of the Forces of Occupation to adopt any measures which he deemed necessary for the requirements of the Armies of Occupation, as well as for the vital interests of the civilian population of the Rhenish occupied territory. On January 31, as a matter of fact, the Commander-in-Chief of the Allied Armies of Occupation issued an order directing the railway personnel in the occupied area to obey the orders of the Interallied Railway Commission and its organs, whereupon the Chairman of the

---

[1]A.R. in O.G., 1920-21, Vol. I, p. 271

National Union of Railway Employees at Cologne announced that the strike would not affect the occupied territory.

In late March and early April some labor trouble was experienced at Neuwied and Bendorf, the most important manufacturing centres of the American area. The Neuwied metal workers won a strike lasting about ten days, a considerable increase in wages being granted them. This dispute was followed by a more serious conflict between the Bendorf metal workers and their employers at the beginning of April. The underlying causes of this unique movement were many, but those of a political nature apparently predominated. The owners tried to take advantage of the fact that certain public utilities were served by their factories; they hoped for a display of armed force by the occupying authorities, but in this they were disappointed. Prominent labor leaders declared that the fairness shown by the local representative of the Interallied Rhineland High Commission contributed much to the final settlement of the conflict.

<u>New Coblenz Kreis President.</u>

Dr. Weil, of the Prussian Ministry of the Interior, a member of the Centrum Party, was appointed administrator of the rural <u>Kreis</u> of Coblenz in January, vice Count Adelmann, who had been appointed President of the government district of Cologne.

<u>Reduction of A.F.G.</u>

With the departure from Antwerp on January 5 of the U.S.A.T. "Cambrai", carrying 1,100 soldiers, the A.F.G. was brought down to its authorised strength of 5,500. When General

Allen[1] assumed command, the forces numbered 75,000. Although the number of troops had been reduced greatly, the limits of the American zone remained unchanged.

Rome Visit.

The Commanding General, A.F.G., paid an official visit to Rome in January and bestowed the Congressional Medal of Honor on the Italian "unknown soldier". He was accompanied by a composite company selected from the American Forces in Germany, under command of Major d'Alery Fechet. Italian troops were turned out at every important Italian stop en route to Rome and the ceremony attending the Commanding General's arrival in that city was truly remarkable. The Minister of War, General Diaz, various Corps and Division commanders, detachments of many different units and large crowds of Italians and Americans were at the station and heard the formal address of welcome made by the Mayor, to which General Allen responded.

The reception accorded the American troops throughout their stay at Rome was sympathetic and enthusiastic. The General called on the Minister of War, the Chief of Staff and General Diaz, and was greeted at the American Embassy by Ambassador Childs and his entire staff.

General de Luca called on him on the 17th and brought decorations from the Italian government for each officer in the party. The General received the "grand cordon of the Order of the Crown". Later in the same day he was conducted to the

---

1 A.R. in O.G., 1920-21, Vol. I. page 255.

Quirinal, where he had an audience of more than half an hour with the King. The outstanding social event of the Rome visit was a luncheon given General Allen and his staff by the King.

The ceremony attending the bestowal of the Congressional Medal was on the 18th in the presence of the King, cabinet members, the American Ambassador, large bodies of Italian troops, the Composite Company, A.F.G., and enormous crowds of people. After the Composite Company had taken its station and a large wreath had been placed at the foot of the altar by two American soldiers, the American Ambassador made an address, at the end of which he requested General Allen to perform the ceremony of homage. The Commanding General made a short address, after which his aide read the Congressional citation. General Allen then laid the medal at the foot of the big golden cross on the altar of the Fatherland where the Italian "unknown soldier" lies. The Prime Minister of Italy made an address and the ceremonies came to an end.

General Allen and twenty-three of the war service members of the Composite Company were decorated by the King with the War Cross on the 19th. The ceremony took place in the court of the Quirinal in the presence of the Queen, the Crown Prince and high government officials.

General Allen returned to Coblenz on January 30th. Accompanied by Major General H.C. Hale and Brigadier General W. H. Johnston, who commanded the First Brigade of the A.F.G., he reviewed and inspected the First Brigade on the same day. The 5th and 8th Infantry Regiments participated, the Brigade presenting a very fine appearance.

Further Reduction of A.F.G.

Orders were received from the War Department on February 17th directing a reduction of the commissioned and enlisted strength of the A.F.G. by 203 officers and 3,000 enlisted men, who were to be returned to the United States by the first transportation available. This decision was favorably commented upon in the German press. The reduction, it was said, would decrease the burdens of occupation considerably, and the action of the United States government was duly appreciated, not only in the occupied area but thoughtout Germany.

In talking with General Degoutte, the Commanding General agreed to the former's request to allow a small number of French troops to be stationed in the American zone, provided that the limits of the American zone be not impaired, that the French troops be under General Allen's command and that the city of Coblenz be off limits to them. The number of troops to come into the zone was to be limited to two or three battalions; they were to take station at Remagen, Mayen, Engers and Neuwied and be allowed to visit Coblenz on special permission only.

Washington's Birthday.

February 22nd, Washington's birthday, was fittingly observed throughout the entire American bridgehead.

Reviewsand Inspections.

A review of the First Brigade, many members of which were under orders to return to the United States, was held on the last day of February. The troops were inspected by General Allen and his staff after the review.

Another review of the various organisations which were soon to return to the United States was held early in March. Headed by the 8th Infantry band, these troops, as they were reviewed by the Commanding General in the Neustadt, presented an admirable appearance.

A series of competitive drills to determine the best drilled company in the regiment was held by the rifle companies of the 8th Infantry during March. The three battalion honor companies met in final competition on Clemensplatz, Coblenz, on March 16. Company "K" was awarded first place. The Commanding General, General McIntyre, General Johnston and many other Allied and American officers witnessed this remarkable exhibition.

Commanding General's Visit to Luxemburg.

General Allen went to Luxemburg on March 21 in answer to an invitation from the Grand Ducal Court.

Report of Return of All Troops.

A cable arrived on March 22, announcing the President's decision to return all American troops to the United States before July 1, and requesting the Commanding General to submit a schedule of transports. The cable contained nothing about the continuance or non-continuance of the Commanding General's status on the I.A.R.H.C.

On hearing the news, M. Tirard[1] left for Paris to do what he could to have his government change Washington's de-

---

[1] A.R. in O.G., 1920-21, Vol. I, page 268.

cision. Lord Kilmarnock[2] and Colonel Ryan,[3] too, took up the question with their government and made an effort to have London influence Washington to reconsider the decision. It was learned that the Belgian government also had taken up the question with Washington.

On the 28th, the Commanding General had a long conference at Paris with Ambassador Herrick in reference to the withdrawal of the A.F.G. and the probable results in the Rhineland of such action. He also called on Lord Hardings, British Ambassador at Paris, with whom he discussed the general European situation and conditions in the Rhineland.

Before he left Paris on the 29th, he learned that Ambassador Herrick had sent a very forceful cablegram to the State Department asking for the retention of General Allen and some forces on the Rhine. The General also learned that the French government had made no official representations against the Washington decision.

---

[2] A.R. in O.G., 1920-21, Vol. I, page 267.

[3] A.R. in O.G., 1920-21, Vol. I, page 267.

Second Quarter of 1922.

German Reply to Reparations Commission.

In its reply of April 9th to the Reparations Commission's note of March 21st, the German government followed the general course of the Chancellor's Reichstag speech.[1] It stated that the situation had become even more acute since the receipt of the note, as a result of continued depreciation of the mark, and predicted starvation and chaotic conditions in Germany that would affect all of Europe unless ways and means were found to prevent further decline of the rate of German exchange. It requested the Reparations Commission to reconsider its decisions of March 21st, and expressed the view that the necessary money for paying the reparations obligations could be raised by a foreign loan only. With regard to the question of guarantees, it contended that the levying of additional taxes was impossible; further, that the financial control outlined was contradictory to former promises made by the Allies. It pointed out again that the taxation compromise constituted the utmost limit of Germany's paying capacity. The government expressed its willingness to explain the financial and economic situation in Germany to the Reparations Commission in further negotiations, as well as the measures adopted for the purpose of improving German finan-

---

[1] See page 11.

ces, but declared emphatically that no German government would ever concede any far-reaching influence on German legislation to foreigners. With the exception of the radical newspapers of the Right and Left, the German press approved the note fully.

Reparations Commission's Answer.

The Reparations Commission's answer to the note was handed to the German War Burdens Commission at Paris on April 13th. The Commission insisted on its original demands for additional taxation and construed the reply as a flat refusal to make any effective effort to levy extra taxes. While a door was left open for further discussions, it was made clear that none of the conditions outlined in the note of March 21st would be annulled or revoked. The German press laid stress upon the sharp tone of the note and upon its full maintenance of the original demands, and the belief was expressed by many that more sanctions would be applied by the Allies in the course of June.

Germany's Conciliatory Attitude.

Germany expected the Allies to enforce coercive measures on May 31 because of its failure to comply with certain demands of the Reparations Commission[1] by that date. However, the danger grew less acute as a result of a German note delivered to the Commission on May 10. Although, in view of the heavy increases already made in taxation, this note pointed to the impossibility of levying still further taxes to the amount of

---

[1] See page 11

60,000,000,000 paper marks, yet it stated the government's willingness to submit a complete scheme for balancing the national budget and guaranteeing future reparations payments to the Commission before May 21st. However, in view of the nation's economic situation and its financial obligations abroad, it stated that it would be impossible for Germany to make the heavy payments due in foreign currencies without the aid of foreign credits. The acceptance of the Reparations Commission's demand for financial control was rendered conditional by recalling the Commission's statement that it "recognises the sovereignty of the Berlin government in questions of expenditure, taxation and financial policy." However, it was added that Germany was ready to "afford all the facilities for investigation asked for by the Commission." In conclusion, the government requested that the time limits laid down in the Commission's note of March 21st be extended because of the absence from Germany of several members of the cabinet by reason of the Genoa conference.

Paris Negotiations.

As soon as it became known that the Reparations Commission had taken the view that the German reply permitted an exchange of opinions, National Minister of Finance Dr. Hermes and State Secretary Bergmann proceeded to Paris. General interest throughout Germany during the last two weeks of May was centered on the negotiations carried on there, whereby the conditions were determined under which a second moratorium was finally granted on May 31st, after the acceptance of the Reparations

Commission's demands on the part of the German government. These demands required a restriction of public expenditures, the fixing of the floating debt as of March 31st, an investigation of German finances, the prevention of unauthorised exportation of capital, the autonomy of the Reichsbank and the publication of financial and economic statistics. The acceptance of these terms was, however, based on a proviso that a foreign loan be granted Germany. It was dispatched to Paris on May 29th without the consent or knowledge of the Reichstag, which body was subsequently confronted by a fait accompli.

The Reparations Commission found the answer satisfactory in general, although it required certain explanations. This settlement, for the time being, obviated any coercive measures on the part of the Allies such as had been expected on May 31st, the so-called "critical day".

Genoa Conference.

The much-heralded assembly of delegates of European nations, including Soviet Russia, was formally opened at Genoa on April 10th. A German delegation was present. This included Chancellor Wirth, Foreign Minister Rathenau and the Ministers of Economics and Finance. The principal purpose of this conference was a discussion of the general economic conditions of Europe and the world and the adoption of adequate measures for stabilising those conditions and repairing the damages of war. The Versailles treaty or any other agreement concluded by the Allies on French soil was not to be endangered.

The ceremonial opening of the conference took place

on April 10th, Premier de Facta of Italy being unanimously elected president. In summarising the conditions to be fulfilled by the assembly, Mr. Lloyd George stated that the main one was a recognition of the principles of civilised nations on the part of all the nations participating therein. He said that no country which had assumed obligations toward another one might withdraw from these obligations. Chancellor Wirth declared that equality of rights of all the nations represented at the conference was a matter of urgent necessity. He signified Germany's sincere willingness to cooperate towards the success of the conference and said that the distrust still prevailing in the political atmosphere, as well as the tension between nations, must be removed. The restoration of international commerce was, he said, the most important condition for a gradual recovery of Europe.

Foreign Minister Rathenau delivered an address on Europe's general economic situation at another meeting. He touched upon the reduction of work and purchasing power and the unemployment problem. The German Foreign Minister suggested a combination of financial and economic measures, including stabilisation of exchange rates and freedom of trade, for the purpose of improving the situation.

Rapallo Treaty.

The news of the conclusion of a Russo-German treaty at Rapallo on April 16th came as a surprise to the delegates. Although nominally of an economic character, Allied circles looked upon this act as a highly political one and as a possible first step towards a Russo-German alliance. The far-reaching possibilities of the pact were duly recognised by the entire German press from the extreme Right to the radical left.

The indignation and surprise of the Allies and of the Little Entente were expressed in a joint note of nine powers which was delivered to the Germans on April 18th. After charging the Germans with bad faith and condemning the secretly negotiated treaty, the note requested them to withdraw from the conference committee on Russian affaris. A crisis and an abrupt end of the conference were avoided only by direct negotiations between the Allied and the German delegates, which resulted in a German note of acceptance of the Allied viewpoint. Germany withdrew from the joint deliberation on the Russian problem, "inasmuch as questions would be under discussion which were already settled for Germany by the Russo-German pact."

Notwithstanding the fact that the majority of the Allied delegates considered the German reply satisfactory, M. Barthou, Chairman of the French delegation, stated in a letter to the Italian Prime Minister that the Russo-German treaty was a violation of the Cannes program. He referred to the "mendacious statements" of the Germans, which expression was strongly repudiated by Dr. Wirth.

Thanks to Mr. Lloyd George's energetic attitude, the situation was saved at a meeting of the delegates of the Allies and those of the Little Entente held on April 22nd. This meeting drafted a note to the Germans, delivered the next day, in which it reserved the right of the interested governments to declare null and void all the terms of the Russo-German agreement that might be considered to be in violation of existing treaties. It declared the treaty incident closed. The Germans, therefore, did not reply.

Conclusion of Genoa Conference.

The Genoa conference came to a close on May 19th. At its last plenary session, German Foreign Minister Rathenau spoke on its achievements and the prospects of an improvement in the European situation. He maintained that this improvement would make itself felt gradually, but said that certain conditions must be fulfilled which had been hardly touched upon in the course of the deliberation. First of all, he said, the reconstruction of Europe was impossible without the active participation of the United States. He then pointed to the necessity of restoring full confidence among the European nations. He declared that "public opinion in certain countries must be demobilised", since Europe was still in a state analogous to that of war, and added that reconstruction could not be accomplished by coercive measures. He said that the first step towards a rapprochement between the western countries and Russia, which had been taken at Genoa, lent a historical significance to the conference.

The Reparations Commission sent another note to the German government on May 4th as a result of the Rapallo treaty. It reserved to itself the right to examine the effects of the treaty from time to time with a view to protecting the Commission's prerogatives and Allied interests. At a reception given by German press correspondents at Genoa and in response to certain statements in the foreign press, Chancellor Wirth declared emphatically that the Russo-German treaty had been published in its entirety and that the stories of a supplementary military pact were pure invention.

Conference of Experts scheduled for Hague.

Although a temporary disagreement among the Allies threatened the success of the conference for a time, Mr. Lloyd George managed to reestablish sufficient concord among them to be able to persuade them to consent to further deliberations to be held by experts at The Hague on June 26th. Germany was not to be invited to attend, since the deliberation there were to relate solely to the Russian question which, as far as Germany was concerned, had, as stated, been settled by the Rapallo treaty. In summarising the results of the Genoa conference from a German viewpoint, the German press expressed the opinion that it could hardly be considered a success, but that it would be wrong to speak of it as a failure. The German delegation returned to Berlin on May 21st and immediately reported on its activities at several cabinet meetings.

Status of Police Forces.

The German Foreign Office transmitted a note to General Nollet, Chairman of the Interallied Military Control Commission, on April 5th, replying to his note of March 23rd. The government recognised the status of 1913 as the basis of organisation of the police forces of all the federal states and expressed a hope for a mutual understanding with regard to the other demands.

Upper Silesia.

The German-Polish negotiations on the Upper Silesian liquidation issue, which had threatened to fail several times, finally bore results. An agreement was signed on April 13th, according to which German industries situated in the ceded portion of Upper Silesia should not be interfered with by the Poles during the 15-year period of transition, and property of German nationals should not be subject to liquidation.

United States and Germany.

Full diplomatic relations between the United States and Germany were resumed upon the arrival at Berlin of Mr. Alanson Bigelow Houghton, Ambassador and Plenipotentiary Extraordinary of the United States. He was received by President Ebert on April 22nd, on which occasion very cordial speeches were exchanged.

Reichstag.

After completing the third reading of the taxation bill, the Reichstag accepted the entire taxation program on April 4th, thereby concluding the most important legislation of the session. Minister of Finance Hermes thanked the parties concerned for their cooperation in the successful settlement of the taxation compromise which, he said, constituted the utmost limit of Germany's capacity. The House adjourned on April 17th.

After reopening on May 10th, it took up the postal and railway budgets and the budget of the Ministry of Labor. Before entering upon discussions on the Genoa conference and the Rapallo treaty, the House established the autonomy of the Reichsbank, as had been demanded by the Reparations Commission.

Chancellor Wirth on Genoa Conference.

On behalf of the national government, Chancellor Wirth delivered his speech on the Genoa conference before the Reichstag on May 29th. He regretted that the reparations question had not been discussed at Genoa and stated that the fact that the United States had not participated in the meeting had lessened its importance considerably. He thanked Great Britain which, he said, had succeeded in bringing the conference to an end, "despite all attempts at sabotage." In referring to the complicated situation arising from the reparations, the Chancellor attacked the "policy of time limits" and said that the continuation of sanctions was a negation of the spirit of Genoa.

Government's Clash with Reichstag.

The conclusion of the Rapallo treaty received the general approval of all the parties during the ensuing discussion of the Chancellor's speech. However, the government's approval of Dr. Hermes' reparations negotiations at Paris, and its failure to inform the Reichstag accordingly, were strongly objected to by the parties of the Right, the G.N.P.P. even introducing a vote of lack of confidence. Although, against the votes of the Communists, the G.P.P. and the G.N.P.P., the Reichstag rejected this motion, certain coalition parties were not at all in favor of the treatment the Reichstag had received from the government in connection with the latest reparations developments.

German-Polish Agreement Accepted.

German flags were flown at half-mast on the Reichstag Building on May 30th at a sign of mourning because of the German-Polish agreement on Upper Silesia, which was on the agenda of the session. National Commissioner Schiffer, who signed the agreement on behalf of Germany, recommended its adoption, "in spite of its innate defects." The agreement was accepted against the votes of the German Nationals, the G.P.P., and the Communists, with the reservation that the Supreme Council was in error when it awarded the lost portion to Poland and that, consequently, the decision would not be recognised legally.

The Reichstag adjourned on May 31st.

Riots in Upper Silesia.

In view of the approaching division of Upper Silesia between Germany and Poland, riots broke out towards the end of May in the districts affected, especially in the areas which were soon to become Polish. The situation was very grave until martial law was proclaimed by the Allied military authorities in the districts of Kattowitz, Gleiwitz, Rybnik, Hindenburg and others, after which things improved somewhat.

Another Note on Reparations.

The Reparations Commission transmitted another note to the German government on June 14th in view of the situation created by the failure of the loan negotiations. In referring to the forthcoming deliberations between the Guarantee Committee and the German authorities, the note took up in detail the autonomy of the Reichsbank and the balancing of the German budget, and requested the German government to make every effort to stop inflation.

Guarantee Committee at Berlin.

Chancellor Wirth received the members of the Guarantee Committee officially on June 20th, they having arrived at Berlin on June 18th. He called their attention to the serious apprehensions of the German government concerning another great drop of the mark, which was threatening, and which would upset all calculations made for the reorganisation of German finances. The first official conference between the members of the Committee and a German delegation, headed by National Minister of Finance Dr, Hermes, was held on June 21st.

Reichstag in June.

The Reichstag was busily engaged during June on the bill providing for partial government control of grain and on two interpellations, one of which concerned the Paris negotiations of Dr. Hermes with the Reparations Commission and the government's concessions in connection with financial control and deflation, and the other, conditions in the occupied territory, especially a report regarding allied intentions of withdrawing the troops of occupation should the Rhineland be neutralised politically. Foreign Minister Rathenau replied to the latter on June 21st, stating that the government would never consider any concessions whatever in connection with a political change in the Rhineland. The interpellation regarding the Paris negotiations of Dr. Hermes was replied to in a rather unsatisfactory manner by Chancellor Wirth who, in a brief speech, expressed the hope that the deliberations of the International

Loan Committee, which had been postponed, would be taken up again soon. The government crisis on the reparations policy was prevented from becoming acute, however, by the unhappy event which, only a few days later, shocked the whole world.

The Rathenau Murder.

Dr. Rathenau, Minister of Foreign Affairs and perhaps the best statesman that the German Republic had thus far produced, was murdered on June 24th by members of a monarchical and Anti-Semitic secret society. The crime created a storm of indignation throughout the nation. There were wild scenes in the Reichstag, Prussian Diet and in many State assemblies, where the Centrum Party, the Democrats and the three Socialist parties formed a solid phalanx against the parties of the Right, in particular the German Nationals, who were accused of having brought about the crime through their agitation for reestablishment of the monarchy. Under the slogan, "The Republic is in danger", the national government at once proceeded to publish a drastic emergency decree directed primarily against monarchism. This was followed by another manifesto of considerably greater severity a few days later.

Work was suspended for several hours in all large cities on June 28th, and the working-people put forward some very radical demands. Carlsruhe and Darmstadt were mentioned as being the scenes of riots on that day. Another bill for the protection of the Republic, which the government found necessary to propose in addition to the temporary emergency measures already mentioned, was introduced into the Reichstag in this connection."

Opposition in Bavaria.

Bavaria's recalcitrant attitude towards the extraordinary measures taken by the national government was displayed clearly in a speech by Bavarian Prime Minister Court Lerchenfeld on June 28th. In it, he declared that he contested openly the National Minister of the Interior's right to demand the exectution of such measures, and that he considered that move an interference in Bavaria's sovereign rights.

Germany was still in the throes of political upheaval as the month of June and this period closed.

Rapallo Treaty.

The most important economic event of this period was the conclusion of the economic agreement between Germany and Russia, known as the Rapallo treaty, on April 16th at the town near Genoa which gave the treaty its name. This treaty, referred to above, was signed by Herr Rathenau and M. Chicherin and provided that Germany and the Russian Soviet Republic should mutually cancel repayment of expenses and damages caused by the war as a result of military measures, including all requisitions in the enemy country and compensation for damages caused civilians. Russia abandoned all claims to reparations from Germany as guaranteed by Article 116 of the Versailles treaty, and Germany waived all claims resulting from the enforcement of laws and measures of the Soviet Republic which affected German nationals or the rights of the nation, with the reservation that the Soviets were not to grant reparations for similar claims that might be made by other powers. The two governments also agreed that the rights of nationals of either party within the territory of the other, as well as the regulation of trade relations, were to be based on the principle of the "most favored nation," while both powers guaranteed each other the settlement of all questions on the principle of reciprocity. Diplomatic and consular relations between Germany and Russia were to be resumed immediately.

In addition to this agreement with Russia, an economic treaty was concluded with representatives of the Memel District in April and negotiations, with the same end in view, were in progress with Finland and Spain.

Deliveries in Kind.

The reparations question, as usual, managed to hold the centre of the stage throughout this period. On April 2nd the Reparations Commission gave its consent, with certain reservations, to the three agreements on deliveries in kind and to the Wiesbaden regulations of the question of substitution. M. Bemelmans, Belgian delegate, entered into negotiations with the German government and secured the necessary signatures to these agreements, subject to subsequent ratification by the Reichstag. The first of the reparations payments under the new schedule, amounting to slightly over 18,000,000 gold marks, was made on April 12th.

Committee of Experts Appointed.

About this time, the Reparations Commission published an official communique regarding the appointment of an expert committee to investigate conditions for a German foreign loan. The experts were to be empowered to procure further views on details of such a loan from New York, London, Paris, Rome, Brussels, Amsterdam, Berne and Berlin. M. Delacroix, Belgian delegate, was appointed chairman of the committee. Mr. J. Pierpont Morgan accepted the invitation from the Reparations Commission to participate in the meetings, and announced that a German loan in America might meet with some success. This did much to restore German faith in the possibilities of such an action.

German Note of May 29th.

The German note of May 29th, in which the demands made by the Reparations Commission were accepted, contained, in part, the following points:

"The increase in German receipts provided for by the present budget is based upon German taxes, including the taxation compromise, and takes into consideration last year's revenues and the depreciation of German exchange.

"With regard to expenditures, the allowances for the administration of national enterprises have been eliminated. Only 950,000,000 marks were provided for the reduction of food prices for the current year, in comparison with 17,200,000 marks in 1921. All other subventions have been abolished. On the other hand, a saving of 3,000,000,000 marks has been provided for and savings in the national administration are being contemplated. The government has decided to appoint a special commissioner of savings in order to assure execution of these measures.

"The government will endeavor to prevent further increase of the floating debt. However, it is convinced that without sufficient financial aid through a foreign loan, its efforts, under present economic and financial conditions, will be in vain. Provided this loan be floated within a reasonable period of time, the government undertakes to carry out the following measures:

1. The amount of the floating debt on March 31st to be considered the normal maximum debt;

2. If on June 30th or on the last day of one of the ensuing months the floating debt exceed the normal maximum amount, steps will be taken to guarantee repayment of the excesswithin the following three months.

"The government further gives its consent that the Reparations Commission make necessary investigations of German finances, as stated in previous communications. It also agrees fully with the Reparations Commission that the illegal exportation of capital must be stopped and will adopt necessary measures to this end. A program of these measures will be submitted to the Commission before June 30th. The autonomy of the Reichsbank was established by special act of May 26th, 1922. Further, the government has directed that economic and financial statistics be published as they were in pre-war years."

International Loan Committee.

Germany accepted the Reparations Commission's demands and the application of further sanctions was avoided thereby. Interest was thereupon centered in the activities of the International Loan Committee at Paris which, under the chairmanship of Mr. J. Pierpont Morgan, had been seeking a way out of the difficulties of the reparations payments since May 24th. The labors of the Committee proved to be in vain, however, for it was found that the main obstacle — the reparations debt of 132,000,000,000 gold marks claimed by the Allies from Germany - could not be overcome. The Committee decided not to touch upon Germany's paying capacity in its report because authorisation to do so had been given it only by a majority vote of the Reparations Commission. When publishing its decision on June 10th, the Committee declared that a loan was unfeasible because Germany could offer no securities, its entire economic and social existence being endangered. The report stated that the members of the Committee would be ready at any future time to lend their services to the Allied powers. In view of this negative result of the activities of the International Loan Committee, the German press expressed fears that the Reparations Commission would insist on the fulfilment by the German government of its last promises which, under the changed conditions, the government considered invalid.

Fluctuating Exchange Rates.

The Paris negotiations with the Reparations Commission had the effect of influencing the money market most erratically, in addition to which the German rate of exchange faithfully reflected all rumors emanating from the French capital in regard to the success or failure of a possible international loan. The value of the mark, which was 284 to one dollar on April 28th, fell to about 310 on May 3rd, only to rise again to 284 on May 10th. A partial cause of this fluctuation was seen in the issue of Reichsbank and Darlehenskassen notes to the value of over 8,000,000,000 marks in the week ending April 29th, and in the increase of the floating debt by over 4,000,000,000 marks during the last ten-day period of April. After May 10th, the Paris negotiations caused the mark to drop as low as 317 to the dollar, which point it reached on May 22nd. Towards the end of the month, however, when optimism as to a loan was at its height, it again reached 286.

The rate of exchange of the mark continued to be influenced during the first part of June by the deliberations of the Loan Committee. Hopes that a loan would still be granted, despite apparent difficulties, held the mark at a rate in the vicinity of 290 to the dollar until June 9th, when the failure to agree became known and the mark fell rapidly, reaching 320 on the 12th. How precarious the general economic situation in Germany had become was indicated by the unprecedented depreciation of the exchange rate which, beginning on the day of the Rathenau murder, sent the mark tumbling down from one low level to another until the record quotation of 445 was reached. Its drop received an additional impetus in the heated political strife which followed the murder of the Foreign Minister. The obscurity of the internal political situation was rapidly destroying the confidence of foreign business which, with Holland in the lead, took steps to withdraw its credits from Germany. The Reichsbank threw enormous sums in foreign securities on to the market in the last days of June in a great effort to stem the tide, but even these measures only temporarily stayed the downward trend of the mark.

National Finances.

The total national receipts during the second ten-day period of May, in respect to customs duties, taxes, etc., amounted to 3,946,102,000 marks. Public services made a profit of 1,763,092,000 marks on running expenses, of which the postal service contributed 464,634,000 marks and the railways 1,303,453,000 marks. The floating debt increased by 3,992,570,000, thus bringing it up to 285, 682,789,000 marks. This amount was increased to 289,246,065,000 marks in the third ten-day period of May.

Hamburg Harbor Traffic.

The traffic of the port of Hamburg for May, for the

first time since the war, not only equalled but exceeded the peace-time tonnage of 1913. The arrivals aggregated 1,143 boats, totalling 1,243,815 net registered tons, as compared with a total tonnage of 1,193,526 in 1913. Departures were 1,449 boats, totalling 1,325,714 net registered tons, as compared with 1,362 boats aggregating 1,253,987 tons in May, 1913.

Reestablishment of German Merchant Marine.

Signs of the reestablishment of the German merchant marine continued to be noticed in June. The shipyards of the country were busy and the launching of vessels was reported regularly, the largest among these vessels being the 35,000 ton twin-screw steamer "Columbus". The Free Cities of Hamburg and Bremen undertook extensive harbor improvements in anticipation of the trade which the future seemed to promise.

Coal Deliveries.

The intense excitement on the money market was instrumental in rendering other phases of German economics acute, such as the coal situation, the food question, etc. The Reparations Commission demanded delivery of the coal which was due the Allies, and since Germany had fallen in arrears with its delivery of coke especially, a demand was submitted to make up the shortages, whereby the July coke deliveries alone would be brought up to a total of 900,000 tons. The first effect of this demand was a 10% cut in the coal allowances of the smelting works of the country at a time when a large number of furnaces had already been extinguished on account of scarcity of fuel. The real trouble, however, seemed to be inadequate production, and attempts were made to introduce extra shifts in order to remedy the evil. Those, however, the miners refused to work.

Cost of Living.

Prices and the cost of living continued to increase during May and June. As a natural consequence, incomes which had seemed ample only a short time before proved to be entirely inadequate, so that demands for higher wages were still the order of the day, although general conditions favored all industries and there was no unemployment to speak of.- The inevitable wage disputes led to strikes over practically all Germany, the biggest ones being recorded at Berlin and Frankfort. These drastic measures on the part of the workers were resorted to when higher incomes could be obtained by no other means.

In order to show what the Coblenz "Hausfrau" had to contend with during this period, the prices of various articles, picked at random, are given below, with varying dates for the sake of comparison:

(In marks)

	Spring 1921	April 10, 1922	June 30, 1923
Potatoes, per lb.	0.65	3.40	3.30
Margarine, " "	10.-	33.-	41.-
Butter, " "	30.-	75.-	85.-
Eggs, each " "	1.50	4.80	6.80
Beef, " "	13.-	44.-	60.-
Pork, " "	17.50	46.-	72.-

In Occupied Zone.

Separatism.

The anti-Prussian activities of the Rhenish Separatists,

especially of the Smeets group, were prominent in various parts of the occupied area during this period. The districts of Mayence, Treves, Bonn and Aix-la-Chapelle were the principal centres of the movement. Herr Smeets surprised the Rhineland with an open letter to the Interallied Rhineland High Commission, published in the "Rheinische Republik", in which he demanded that "all Prussian officials, especially the German Rhineland Commissioner and his staff, be expelled from the area and replaced by true Rhinelanders," and that the Commission suspend the circulation of a great number of prominent Rhenish newspapers of all party shades, because of their "agitation against the Rhenish movement."

Opposition to Separatism.

On the other hand, the Rhenish political parties, including the Independent Socialists and Communists, and the Free and Christian Trade Unions and other organisations, united in a common effort to fight the Separatist movement. Resolutions protesting against Separatist machinations were passed in a large number of meetings, and it soon appeared that the overwhelming majority of the inhabitants of the Rhineland were strongly opposed to a change of their political status in relation to Prussia and Germany.

As a result, the agitation of the Rhenish Republicans subsided somewhat in June. In addition, the Separatist press was beginning to meet with difficulties, the entire technical staff of the printing office at Bonn, for instance, which published the "Rheinische Republik", (Herr Smeets' paper), deciding to have no further hand in the printing or publishing of that paper.

The climax of the anti-Separatist movement in the Rhineland was reached at a big mass meeting held by all political parties at Honnef on June 6th, when Herr Otto Braun, Premier of Prussia and a Majority Socialist, attacked the Separatists strongly in a remarkable address.

Herr Smeets before Court again.

Herr Smeets once more appeared before a Cologne court on June 12th to answer for some statements made in certain articles which had appeared in the "Rheinische Republik" and which were regarded as an insult to President Ebert. He did not succeed in clearing himself, and the court, by combining the verdict with a former five months' sentence, passed a total penalty of seven months' confinement on the Separatist leader. At another trial on June 19th, the court sentenced him to two months' more confinement on a charge of insulting the Cologne police. The execution of these sentences was withheld, pending a decision by the interallied Rhineland High Commission, which, it will be remembered, had allowed Herr Smeets to be tried but reserved the right to revise any sentence passed on him.

In American Area.

Separatism.

The most important feature of the political life in the American area during this period was the activity shown by the Smeets group of Separatists, previously little known in the Coblenz district, wereas Dr. Dorten had almost passed out of the public eye. Although the area was flooded with propaganda matter by the Smeets people, it cannot be stated that Separatism made any headway at Coblenz. On the contrary, the opposition of the anti-Separatists seemed stronger than ever. A concert given in front of the Royal Palace on May 18th on the occasion of a Duerer festival resulted in a spontaneous demonstration by the Coblenz population, headed by Chief Burgomaster Russell, against the movement.

May Day.

The Coblenz group of Majority Socialists, reinforced by the Independents, celebrated May 1st by a joint demonstration which culminated in a meeting at the Festhalle (Civic Amusement Centre). Work was suspended in the majority of the local ships and factories. The occasion was also observed by the workingmen of the towns and villages throughout the area. The day, however, passed quietly, without untoward incidents.

Death of Ober-Praesident of Rhine Province.

Herr von Groote, Ober-Praesident of the Rhine Province, died at Maria Laach in the American area as a result of heart failure on May 10th. In reviewing his career, the local press praised his rare tact and the efficient manner in which he had managed the affaris of his office. His funeral at Neuwied was attended by representatives of the Interallied Rhineland High Commission and the forces of occupation.

Political Activities.

The various political parties represented in the area held the usual number of meetings and conventions during this period. An important meeting was held by the German People's Party at Cochem on May 27th. It was attended by its district delegates and local delegates of the electoral districts of Coblenz and Treves. The meeting was typical of most of the political gatherings of that time, in that a resolution was passed expressing the loyalty of the Rhenish population to the nation and Prussia. The question of filling the vacant post of Ober-Praesident of the Rhine Province, which had aroused much interest among the various parties, was also discussed. A resolution was passed calling for the appointment of a man who was a confirmed anti-Separatist, aside from which his experience and ability only should be considered.

Socialists Protest against Rathenau Murder.

The murder of Foreign Minister Dr. Rathenau late in June stirred the population of the American area to intense excitement. Meetings were held at Coblenz and elsewhere by Socialists, Democrats and Centrists. The Socialists met in the Festhalle at Coblenz on June 28th. The speaker of the day, after praising the late Minister's achievements, blamed German National circles for the crime. He criticised the attitude of the German People's Party local organ, the "Coblenzer Zeitung", for its attitude towards the murder and expressed the view that Coblenz was still a hot bed of reactionaries.

The new German black, red and gold colors were seen for the first time at Coblenz on the occasion of this meeting.

Centrum and Democratic Meetings.

Although not participating in the Socialist demonstrations, the Centrum and the Democratic parties, on June 29th and 30th respectively, held dignified meetings in honor and memory of the murdered Minister. The programs of these gatherings were restricted to a review of the life and activities of Walter Rathenau and closed with a solemn oath, taken by all present, to stand united in a joint defence against the unscrupulous agitation of all anti-constitutional elements .

The various Summer Solstice celebrations, which had been widely advertised by the parties of the Right, were held on June 24th in Coblenz and vicinity and passed off without disturbances. Under the immediate impression of the Rathenau murder, all the speeches were couched in rather cautious terms.

Visit of Bishop of Treves.

Catholic Coblenz celebrated the festival of Corpus Christi on June 15th by a large procession and a special open air mass. The newly elected Bishop of Treves arrived in the evening and was given a great ovation by the Catholic population, which also attended a special mass celebrated by him at the Church of St. Castor. From a political point of view the Bishop's visit was a powerful demonstration of Catholicism and proved the undisputed predominance of the Roman Church and its political adherents — the Centrists —

in the American area.

Economic Outlook.

The economic outlook in the American area at the beginning of this period seemed rather favorable. Practically all of the local industries were busy and unemployment figures, already so low as to be negligible, were steadily decreasing. The only disturbing feature was the pessimistic forecasts of crop experts, who predicted failure for practically all crops except potatoes, in which, from a producing point of view, the rural districts of the American area were only slightly interested.

Strikes.

Following the comparative calm that had reigned in the world of labor during April and May, the month of June witnessed several strikes of more or less serious nature. The most important of these was a strike of the Coblenz shop-clerks and employees, which affected practically every business-house of note in the city. The employees demanded chiefly a 40% increase in wages, which was finally granted them. Upon request of the trade union leaders, the Interallied Rhineland High Commission gave its consent to a strike not only to the clerks, but also to the brewery employees and tramway-men. These disputes had hardly been settled in favor of the strikers, when the Coblenz Transport Workers' Union (draymen, teamsters, etc.), evidently encouraged by the success of the others, proclaimed a strike on June 12th. This lasted only five days, all firms then agreeing to grant the men's demands.

Last, but not least, the Woodworkers' Union of Coblenz followed suit, proclaiming and winning a strike in quick order.

Matters Pertaining to Withdrawal of A.F.G.

General Allen formally notified General Degoutte on April 4th that orders from the United States required complete evacuation of the American Forces by July 1st. The 1st Brigade being about to be demobilised, General Allen requested that its commanding officer, Brig. General W.H. Johnston, be allowed to go to Mayence to take up with French headquarters the question of replacing the American troops in the Coblenz area by French troops. The 1st Brigade was demobilised on April 9th and Brigadier General Johnston was appointed general liaison officer with the French Army of the Rhine in compliance with War Department instructions.

M. Tirard told General Allen on April 6th that the French government had made representations at Washington, as M. Tirard had asked, requesting that not all the American troops be withdrawn.

Ex-Governor Lowden of Illinois arrived at Coblenz on April 15th. After conferring with several of the Commissioners and Deputy Commissioners, he sent a cablegram to Secretary Weeks recommending the retention of General Allen and a sufficient number of troops on the Rhine to at least guard the flag for some time, stating that such action was indispensable if the strained relations between the French and Germans were to grow no worse.

Ambassador Houghton arrived from London on April

19th and then left for Berlin to take up his new duties there. He had talks at Coblenz with the Commissioners and Mr. Dressel, who had come from Berlin to meet him. After several conferences with General Allen's colleagues regarding the situation on the Rhine, he proposed to cable to Washington concerning the projected evacuation. He also informed General Allen that Ambassador Harvey was making every effort to prevent the American representation on the Rhine from being recalled.

German Reichs-Kommissar Prince Hatzfeldt called on the Commanding General on April 20th and was much disappointed when informed of the United States Government's decision to remove the flag and the entire American representation from the Rhine. He was of the opinion that the decision was inopportune for Germany and that our departure meant the end of the office of National Commissioner.

A code cable arrived on May 2nd, delaying the sailing of the 8th Infantry, less one battalion, until July 20th, on account of complications in quartering the troops in the United States. This caused belief that it was the President's intention to maintain some troops on the Rhine, since no difficulty had ever been experienced in quartering all the troops returned from the A.F.F.

General Degoutte sent General Allen a tentative plan to govern the French troops in the American zone under General Allen, recognising the latter's authority as "territorial commander". It was answered at once by General Allen, saying that it was acceptable with the understanding that regulations then in force with reference to undesirable women and drink stronger than light wines and beers should apply to French as well as American soldiers, and that the American provost courts should continue to exercise jurisdiction over all cases between the German population and military in the American zone.

The 4th French Dragoons Brigade, the first contingent of the French Army to be stationed in the American area, arrived on May 5th. The Commanding General of the Brigade, Brigadier General de Partouneaux, and his staff, arrived on the same day.

The 2nd Battalion, 8th Infantry, left for Antwerp on May 15th, en route to the United States, where it took up its station at Fort Moultrie, S.C.

The Provisional Cavalry Squadron, A.F.G., was demobilised on May 20th, orders to that effect having been issued by Hqrs.. A.F.G.

1,200 Officers and Men to Remain on Rhine.

On June 3rd, while at a dinner given by Lord Kilmarnock at which all the Commissioners and their deputies were present, General Allen was handed a W.D. cable announcing that a force not to exceed 1,200 officers and men would remain on the Rhine indefinitely. The contents of the cable were made known to the officials of the Commission present, all of whom expressed great satisfaction.

Prince Hatzfeldt called on the Commanding General to express his pleasure at the retention of the A.F.G. on the Rhine, and mentioned it as unusual that the authorities of an occupied country should desire the continued maintainence of a foreign force within its confines.

General Allen Visits Berlin.

While in Berlin from June 15th to 17th, the Commanding General was afforded an excellent opportunity of discussing Rhine questions and matters concerning the general European situation with both Allied and German diplomats and representatives. He learned that the French had represented to England that the attitude of Lord d'Abernon, British Ambassador, was hardly compatible with what was to be expected of the representative in Germany of a country still engaged with France in the execution of the Versailles treaty.

At a dinner given by Colonel Creed Cox, military attache at Berlin, General Allen met and talked with all the chief members of the Allied Board of Control. Although it was admitted by these officials that Germany was "materially" disarmed, they stated that 250 officers were still being maintained who, it appeared, were striving for its "moral" disarmament.

Before leaving Berlin, the Commanding General received all of the principal American journalists at the American Embassy and talked to them for some time in the presence of Mr. Houghton. His statements covered principally matters connected with the Rhine and were so phrased that they might have been made to a chauvinistic German or Frenchman.

The opinions of M. Poincare on General Allen's talk to the American press representatives at Berlin were later given

General Allen by M. Roussellier.[1] General Allen informed M. Roussellier that he had weighed his words carefully and that he had said nothing which could offend either M. Poincare or Herr Rathenau, and he requested M. Roussellier to repeat that to his government.

French Plans in American Area.

Plans were received from French G.H.Q. on June 7th involving the bringing into this zone of a complete infantry division and construction of barracks at Andernach, Neuwied and Sinzig. The Commanding General believed that such a big force and the proposed constructions were not necessary, and informed Captain Hendrickx, liaison officer, that under no circumstances should such a project be carried out in the American area.

New Base Port.

The War Department approved the Commanding General's proposed routing of future supplies via Hamburg, instead of Antwerp, on June 8th, thus realising a considerable economy for both the A.F.G. and the German government.

Miscellaneous.

M. Albert Deveze, Belgian Minister of War, and M. Maginot, French Minister of War and Pensions, accompanied by a number of Allied military chiefs, visited the American area

---

[1] A.R. in O.G., 1920-21, Vol. I, page 269.

as guests of Belgian High Commissioner Baron Rolin-Jacquemyns[1] on April 25th. M. Deveze was decorated with the Distinguished Service Medal by General Allen. The ceremony was simple and took place at the Commanding General's house in the presence of the Minister's suite and Baron Rolin-Jacquemyns. M. Deveze in turn decorated three American officers.

A very successful Interallied Horse Show was held by the A.F.G. at Carnival Island on May 26th and 27th.

May 30th, Memorial Day, was fittingly observed throughout the American area.

General Colagher of the Reparations Commission called on the Commanding General to discuss the participation of the United States at a meeting of the Commission on Army costs. He suggested that such a participation might be unofficial and would have the advantage of preventing a complete ignoring of the American rights, as had happened at Cannes.

Brigadier General W.H. Hay, Chief of Staff, A.F.G., left Europe on the "Cantigny" for reassignment in the United States.

Mr. Charles H. Grasty, of the "New York Times", arrived to see the Commanding General. He wanted General Allen to facilitate his visit to Marshal Hindenburg, from whom he desired to get an authoritative statement concerning the influence of the American troops in deciding the war. He believed that an earnest expression from Marshal Hindenburg would be of great value to history. This appeared in accord with the views of Generals Pershing and Harbord, as well as others who believed

---

[1]A.R. in O.G., 1920-21, Vol. I, page 262.

that the great role played by our troops in the world war had at times been reported in such a way as to minimise our efforts, and that a public statement from Marshal Hindenburg would go far toward giving public opinion a better estimate of what our participation meant. Due to the Field Marshal's hostility to interviews with newspaper correspondents, Mr. Grasty did not succeed in his mission.

Among other prominent visitors to the American area during this period, (practically all of whom were entertained by the Commanding General), were Mr. Ellis Loring Dresel, Charge d'Affaires at Berlin before the appointment of the Ambassador; Major General P.C. Harris, Adjutant General of the Army; Representative John Q. Wilson from Connecticut, who was sent to the American area by President Harding as a special investigator and observer ; Major General James G, Harbord, Deputy Chief of Staff of the U. S. Army ; Ambassador Fletcher, U. S. Ambassador to Belgium ; Major General W. G. Price, Commanding 20th Division, Pennsylvania National Guard ; Major General George P. Scrivens, U.S.A. retired, and Major General Robert L. Bullard, U.S.A.

Third Quarter of 1922.

Socialists versus Bourgeoisie.

In an attempt to secure the passage of the law for the protection of the Republic, which was primarily directed against the Right, the Majority Socialists formed a so-called cooperative working union with the Independent Socialists. Simultaneously, efforts were made to extend the government coalition to the Left to include the Independents, and an announcement was made that the Majority Socialists would not countenance an extension to the Right, for which the German People's Party had signified a desire. Having become alarmed at this decided swing of the political pendulum to the Left since the assassination of Dr. Rathenau, the Centrum Party proposed a smiliar working union between the Bourgeois coalition parties and the German People's Party. Pursuant to advice from the National President, the Majority Socialists receded from their original demands in regard to the bill for the protection of the Republic and persuaded the Independents to follow their example. In consequence, the bill became a law on July 18th in a form least objectionable to the Bourgeois parties. The idea of extending the government either to the Right or Left was abandoned for the time being. The contemplated cooperative association having failed to be formed, an announcement was made by the Centrum Party to the effect that it desired to expand so as to include all Christian voters who favored the existing constitutional government. It appeared, therefore, that a Christian constitutional party with the

Centrum Party as a nucleus represented the aspirations of the Centrists.

Retirement of Monarchist Officials.

The new policy of enforcing republicanism resulted in the retirement throughout Prussia of a number of high government officials with decided monarchist leanings. The vacancies thus created were filled by trustworthy republicans from the ranks of the coalition parties. The new policy created considerable resentment in the Rhineland, where changes were made without the consent of the Provincial Committee, the approval of which, according to the Prussian constitution, is necessary whenever the appointment of higher provincial officials is involved.

Party Politics.

The movement aimed at the consolidation of the Socialist parties assumed more concrete form in August. This seems to have been the underlying reason for a conference which the German People's Party held at Munich and at which the endeavor of the Party's Reichstag group to bring about a working agreement with the Centrum and Democrats was approved. At the same time, the Centrum Party continued its propaganda for a big constitutional party into which, irrespective of faith, all voters willing to support the existing form of government were invited.

Amalgamation of Majority and Independent Socialists.

The outstanding feature in September was the amalgamation of the two Socialist parties, i.e., Majority Socialists and Independents, which was accomplished at Nuremberg on September 24th. The delegates of the two parties confirmed their fusion there, preparations therefor having been made at the conventions at Gera and Augsburg. The new party was called the United Socialist Party of Germany (U.S.P.). The platform showed little change as regards the well-known Socialist principles, the plank providing for "ultimate dictatorship by the proletariat", however, being dropped.

Bavarian Particularism.

Of the many issues confronting the national government, the Bavarian question disturbed Germany's internal equilibrium most during the summer of 1922. The law for the protection of the Republic, enacted by the Reichstag at its closing sessions, became the bone of contention between the national government and Bavaria. This, the largest state of the commonwealth next to Prussia, has always been extremely sensitive in the matter of State rights, and its population being monarchist at heart, it naturally was loath to accept any dictation from the "Socialist Berlin government" in the form of one-sided laws, especially when they were primarily directed against monarchism. The Bavarian government consequently refused to execute the national measures throughout the State, the Palatinate excepted, by resorting to the questionable subterfuge of promulgating an emergency decree of its own on July 24th, whereby the competence of the national authorities was practically repudiated. The national government took the view that this decree constituted a violation of the national constitution and that the departure threatened the unity of the German federation. As a last resort, President Ebert addressed a letter to the Bavarian Premier in an effort to arrive at an amicable settlement. However, the issue was still pending at the end of July.

Compromise between Bavaria and Nation.

The month of August witnessed a considerable quieting of the domestic political situation. President Ebert's letter to Count Lerchenfeld, written in a conciliatory spirit, apparently had the desired effect. The Bavarian government showed a willingness to cooperate and an agreement was reached after a few days of negotiations at Berlin. The Bavarian delegation, headed by Count Lerchenfeld, promised to repeal the Bavarian decree for the protection of the Republic by August 18th at the latest. The issue failed to be settled, however, when Court Lerchenfeld returned to Munich, because the coalition parties, upon whose approval the execution of the protocol signed at the national capital depended, refused to accept the original agreement and demanded a number of modifications referring to State rights and special courts. It consequently became necessary to reopen negotiations with the national government, the results of which were agreed to, despite the dissatisfaction voiced by a certain section of the Berlin press. As the immediate result of the definite agreement, the Bavarian emergency decree for the protection of the Republic was abolished on August 24th.

Agitation in Bavaria continues.

After the settlement of the conflict between Bavaria and the nation, another particularist agitation set in at Munich, which proved that the compromise reached at Berlin on the question of State rights had failed to satisfy the monarchists. The national government apparently made another concession to Bavaria by recalling Count Zech, the national diplomatic agent at Munich, who seemed to have become <u>persona non grata</u> at the Bavarian capital.

<u>Reichstag.</u>

Notwithstanding the disagreement among the parties during the many debates, the Reichstag succeeded in passing the bill providing for a compulsory loan of 70,000,000,000 paper marks, and the national criminal police bill and the amnesty bill, as well. After completing the work on the agenda of the session, the House adjourned on July 18th to reconvene on October 17th, or earlier if necessary.

<u>Constitution Day.</u>

Constitution Day was fittingly celebrated throughout Germany on August 11th, the government authorities setting the pace. The population responded in a half-hearted manner, the dividing line separating monarchism from republicanism being easily discernible. Public demonstrations were held in all the principal cities, at which statesmen, politicians and public functionaries delivered speeches. President Ebert also took occasion to issue a manifesto in commemoration of the adoption of the constitution.

<u>Upper Silesia against Autonomy.</u>

On September 3rd the population of the German portion of Upper Silesia had an opportunity to vote on the question of the autonomy which had been held out to them by the national government during the pre-plebiscite period. The result of the referendum was an overwhelming vote in favor of remaining with Prussia : it was received throughout the country with great satisfaction.

Germany provisionally agrees to Financial Control.

Prior to leaving Berlin for Paris to submit its report on Germany's financial situation to the Reparations Commission, the Guarantee Committee delivered a memorandum to the German government on July 18th, demanding Germany's submission to strict financial control by the Committee. In his reply of July 21st, Chancellor Wirth declared that notwithstanding the severe demands made, the German government agreed to accept the terms for the period during which the proposed moratorium would be in force, provided that the measures interfered in no way with the sovereign rights of the German nation.

Compensation Payments.

In connection with its request for a moratorium covering the reparations payments, the German government forwarded a note to the Reparations Commission and to the interested Allied governments, requesting that the amount of 2,000,000 pounds sterling payable monthly as compensation for losses sustained by Allied subjects during the war, be reduced to 500,000 pounds. In a reply dated July 28th, the French government refused the request and demanded that Germany, within ten days, declare its willingness to continue the monthly payments of 2,000,000 pounds, in default of which certain retaliatory measures would be applied. The German reply of July 31st stated that the cabinet could arrive at no final decision until after all the interested governments had defined their views, and emphasized the necessity of treating the question of a moratorium covering the monthly instalments for reparations and that of the reduction of the compensation payments as one. The French demands, therefore, were not complied with in this note.

German Protest against French Measures.

Germany's reply being considered unsatisfactory, the French government notified the German cabinet on August 7th that certain retaliatory economic measures (measures de rétorsion) had been put into effect on August 5th. These included instructions to the compensation adjustment offices at Paris and Strasbourg to postpone all recognition of German credit until further notice and to make no more payments on account of the German government or from the proceeds of the liquidation of German private property in France. Furthermore, the French government ordered the expulsion of Germans from Alsace-Lorraine. The adoption of these measures caused the German government to address two notes of protest to the French government. It declared that the measures were in absolute opposition to international usage and based upon no article of the peace treaty or the compensation agreement of June 10th, 1921.

Notes on Compensation Payments.

Another note was sent to London, Paris, Rome and Brussels on August 14th, declaring Germany's inability, on account of the unfavorable rate of exchange, to pay the August instalment of 2,000,000 pounds sterling, but stating that one-fourth of that amount had been placed at the disposal of the Allied governments. This resulted in a note from the British government on August 17th. In it, the German Embassy at London was informed that the payment of the 2,000,000 pounds due on August 15th, which Germany was unable to make, would be demanded four weeks from that date, and that steps would be taken by the Allies towards making separate agreements with the German government regarding the future settlement of the compensation payments. This decision was probably the cause of the French government's partial abolition of the measures of retortion in retaliation for Germany's failure to render the compensation payments. In a French note delivered to the German government on August 25th, it was stated that there would be no more expulsions of German nationals from Alsace-Lorraine for the time-being and that the orders directing the sequestration of property in the cases of deportees had been revoked : also, that German credits had been released and that further relaxation of the retaliatory measures was contemplated.

London Conference.

As a result of the urgent character of the reparations problem, the Prime Ministers of France, Great Britain, Italy and Belgium convened at London on August 7th for the purpose of reaching an agreement on the question of granting Germany a moratorium and fixing the terms to be imposed on it if a respite of payments should be conceded. The negotiations came to an end on August 14th without having accomplished anything tangible toward settling the problem, The question of a moratorium was again referred to the Reparations Commission, which indicated in an official communiqué that Germany would receive an early reply on the question of the payment due on August 15th, and that that payment should remain suspended until a final decision was made.

Members of Reparations Commission at Berlin.

After the failure of the London conference, the Reparations Commission despatched Sir John Bradbury, British representative, and M. Mauclere, President of the Guarantee Committee, to Berlin for the purpose of carrying on further negotiations with the German government there. Several proposals to furnish guarantees for a moratorium were submitted by the German representative, and the delegates returned to Paris on August 26th.

Reparations Commission's Decision.

On August 31st, the Reparations Commission decided not to grant the moratorium demanded by the German government and to postpone a final decision pending the execution of certain financial reforms stipulated. At the same time, in lieu of the cash payments due on account of reparations in August, September, October and Decembenber, 1922, it agreed to accept certain treasury bonds running for a period of six months which

Germany was to guarantee on a gold basis.

German-Belgian Negotiations.

Since these bonds were to be turned over to Belgium, that country being entitled to priority, MM. Bemelmans and Delacroix began negotiations with German representatives on behalf of their government at Berlin on September 6th with a view to reaching an agreement on the issue of the bonds and on guarantees whereby they could be immediately disposed of should Germany fail to redeem them at the time agreed upon The deliberations came to an end on September 10th without, however, yielding results satisfactory to the Belgian delegates. The Germans demanded an extension of the six months' period, which the other side was apparently unable to grant.

German-Belgian Issue Settled.

The matter was settled when Dr. Landsberg, German Ambassador at Brussels, informed the Belgian Foreign Minister that the Reichsbank was prepared to sign the bonds in question. The Reichsbank's original request to have the bonds, amounting to 270,000,000 gold marks, run for a period of from twelve to eighteen months having been abandoned, the Belgian government declared that it considered the issue settled on the basis proposed by the Belgian delegation at Berlin. After approval of the agreement, the Reparations Commission turned over the German treasury bonds covering the two reparations instalments due on August 15th and September 15th to the Belgian government on September 26th. The successful consummation of the transaction was primarily due to the efforts of President Havenstein of the Reichsbank, who, after extensive negotiations, the nature of which was kept secret, obtained the cooperation of the Bank of England.

The provisional settlement of the German reparations payments failed to evoke approval in Germany, where expectancy had been keyed up for a moratorium extending over a long period of time. Satisfaction was, however, felt at the fact that a "breathing space" had been obtained.

Agreement between Germany and United States.

An agreement complementary to the German-American peace treaty of August 25th, 1921, was concluded between the German government and that of the United States on August 29th, whereby the latter renounced all claims arising from military pensions or from money allowances paid to American prisoners of war and to families and dependents of men under arms.

Fluctuation of German Exchange Rate.

Economic conditions in Germany during this quarter continued to be influenced by the unstableness of the rate of exchange, which fluctuated in a manner never experienced before. The dollar was quoted in the neighborhood of 500 marks at the beginning of July and fell later to nearly 400 because of the encouraging effect of the vague rumors of a loan and moratorium. Thereafter, the German money market experienced another slump owing to unfavorable comments in the foreign press concerning the moratorium prospects and due to the Reparations Commission's reply insisting upon the payment of the July reprations instalment. The money market then remained rather quiet for a time, with bankers and speculators watching the situation, as a result of which the mark began to strengthen, until more demands for foreign currencies forced the rate down again. This decline exceeded even the most pessimistic predictions when, at the end of July, a rate of 640 to the dollar was reached. Like most of its predecessors, this drop was in the main attributed to the ever-critical reparations problem and especially to the uncertainty of the outcome of the meeting of the British and French Premiers who were to discuss a moratorium for Germany. However, the depreciation was by no means due to external influences alone. The really serious controversy between Bavaria and the national government had much to do with the low state of the mark. Added to this, the general unsatisfactory conditions in practically all other phases of German economics seemed to be undermining the country's credit abroad.

During the early part of August the exchange rate of the mark remained in the vicinity of 750 to the dollar, held there by the hopes arising from the London conference of Allied Ministers, who were once more trying to straighten out the reparations tangle. The rate rose and fell in a limited degree, according to whether it was influenced by favorable or unfavorable rumors from London, until the final failure of the meeting seemed to remove even the last vestige of steadying influence and sent the mark in a panic-like drop down to below 2,000 to the dollar. A reaction during the latter part of the month, however, caused an improvement by over 800 points.

After months of abnormal economic conditions causing violent fluctuations of the exchange rate and unprecedented increase in prices, a period of readjustment and comparative calm seemed to set in in September, and especially in the latter part of the month. The departure of Chancellor Wirth and Minister Hermes from Berlin, so often delayed, seemed to indicate that, for the time-being at least, the foreign political tension had subsided.

This was due, in the main, to the successful completion of Herr Havenstein's mission to London and the consequent satisfactory settlement of the Belgo-German dispute, Although the Reichsbank President, considering the transaction with the Bank of England a private affair, observed strict silence conserning the nature of the agreement arrived at, financial circles in general seemed satisfied with the results obtained. As a matter of fact, the exchange rate of the mark, always considered a reliable criterion of economic conditions in general, remained almost stationary during the last two weeks of September.

Money Stringency.

The ever-increasing activity of the government banknote press and the money situation in general caused businessmen of the country much concern during this quarter. Despite a record-breaking output, the press was unable to overcome the stringency of the money market, so that the national government, in order to alleviate conditions, was compelled to grant many cities and even numerous industrial concerns the privilege of issuing their own emergency notes, especially for the purpose of paying employees' salaries. These measures, needless to say, added to the general inflation and militated against an improvement of the mark.

Credit Crisis.

This scarcity of currency was but a weak reflection of a much greater evil, i.e., the credit crisis. This was sorely felt in practically all branches of German industry and commerce, its cause apparently being an unmistakable reluctance on the part of capitalists to invest their money under existing conditions, except to speculate in stocks or bonds or other papers which promised quick returns. As a result, many industries found it very difficult to bring about an increase in their working capital, so necessary on account of the enormous depreciation of monetary values. Many concerns were consequently compelled to restrict their building activities to a minimum, thereby greatly hampering their efforts at increased production, which seemed to be the only means of a return to normal conditions.

Reichsbank Discount rate Raised.

By raising its discount rate from 7% to 8% on September 21st, the Reichsbank made the third increase of this kind within a few weeks. The enormous rise in prices of all commodities during the preceding weeks was held primarily responsible for this higher rate, inasmuch as the banks, as a result thereof, did not possess sufficient currency to satisfy the demands of commerce and industry.

Fuel Situation.

The coal situation during the first part of this period remained extremely critical. Despite the fact that everything seemed to have been done towards raising the morale of the miners, such as granting them more pay and improving their living conditions, the output of the mines grew smaller from month to month, until the daily average production in the Ruhr Basin in July was below 300,000 tons, as compared with 330,000 in March. The continued fuel shortage obliged German industry to buy British coal, the total import of which for July was estimated to have been more than 1,500,000 tons, or a figure almost equal to the amount of Germany's compulsory deliveries to the Allies. The purchase of British coal was a particular hardship because the steady depreciation of the mark automatically caused the price to rise as far as German consumers were concerned. The American coal strike added another difficulty to the situation, in that the extraordinarily great American demand for British coal also operated to increase the price. In consideration of these conditions, the German government early in the month dispatched a note to the Reparations Commission, in which it invited attention to the greatly diminished German output due to the loss of the Upper Silesian coal fields to Poland and requested a reduction of the monthly compulsory coal deliveries commensurate with the losses in production. It further asked for the privilege of making a part of the deliveries demanded with coal imported from England. The decision of the Reparations Commission, which maintained its demand for compulsory monthly deliveries totalling 1,750,000 tons of German coal and coke, was a great disappointment.

Extra Shift Agreement.

The negotiations concerning extra shifts for the Ruhr miners, carried on for many months, were at last crowned with success when the miners late in August agreed to work three extra shifts a week. This meant not only increased coal production for Germany, but - and this was equally important - it was looked upon by many as the possible beginning of a general movement away from the eight-hour day.

Labor Situation.

In spite of the enormous increase in the cost of living and the almost universal demand for higher wages resulting therefrom, Germany was free from serious labor trouble during August. Although numerous strikes occurred almost constantly throughout the country, none was of a serious nature and most of the disputes were settled by means of arbitration and, in most cases, even without loss of time on the part of labor. This was no doubt due to the fact that the need of more pay was obvious under the abnormal conditions, and to the additional circumstance that practically all industries were busy filling rush orders from abroad and from German retail merchants who realized that every day's delay in replenishing their depleted stocks was bound to bring higher prices.

"Clearance Sales."

Hand in hand with the mark catastrophe went the so-called "clearance sales" in Germany and especially in the

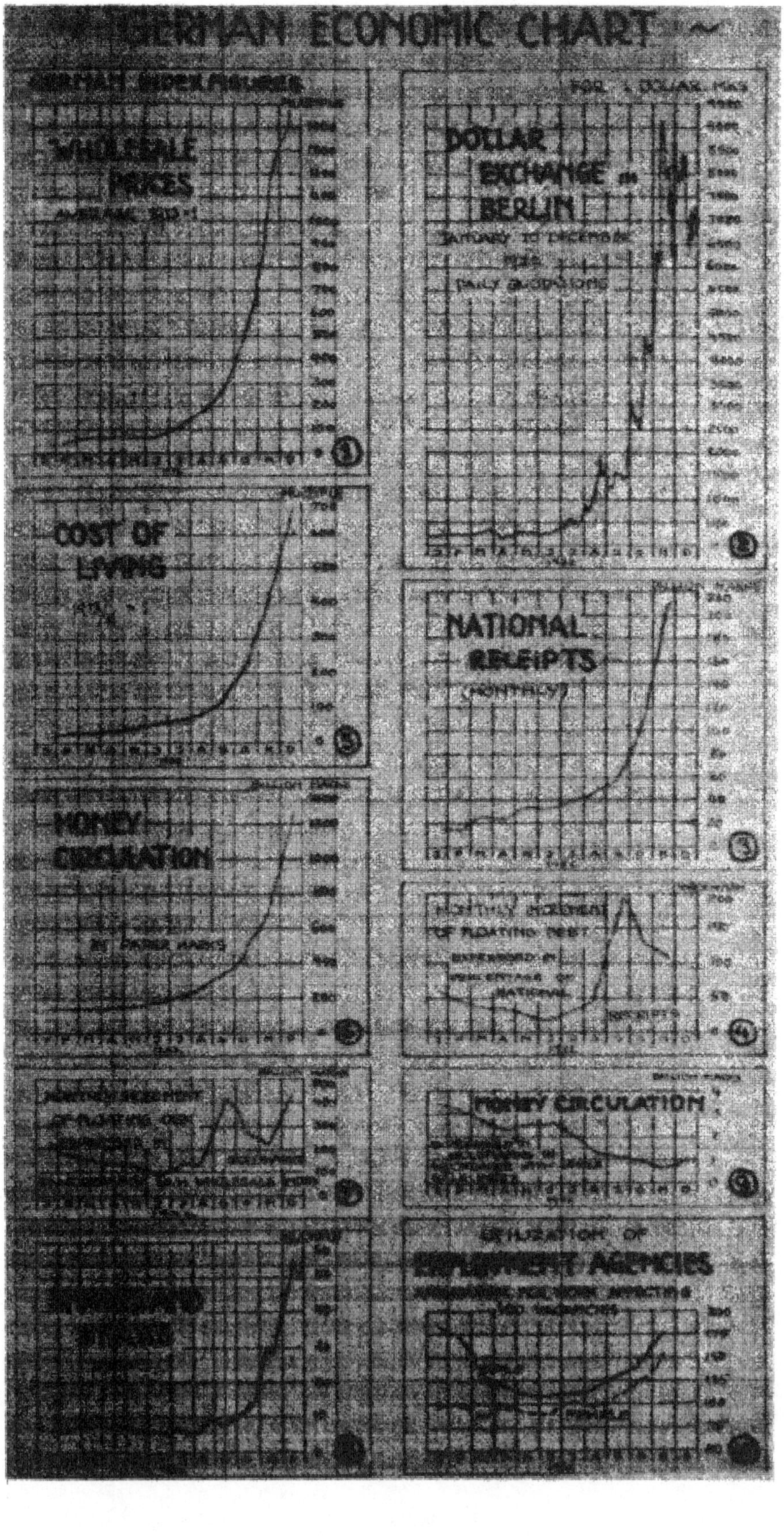
GERMAN ECONOMIC CHART
WHOLESALE PRICES
COST OF LIVING
MONEY CIRCULATION
DOLLAR EXCHANGE IN BERLIN
NATIONAL RECEIPTS
MONEY CIRCULATION
EMPLOYMENT AGENCIES

Rhineland, where foreign buyers taking advantage of the, to them, favorable rates of exchange were very much in evidence. The situation eventually became so bad that merchants decided to charge foreigners a 50% increase on all goods sold them. As a further means of protection, many shops were either closed entirely or were open for a few hours only during the day.

Food Situation.

The government seemed to have the food situation well in hand. With reduced freight rates for potatoes, grain, fruit and other foodstuffs, and with plans all made by the Railway Administration for the uninterrupted transportation of food supplies from the agricultural districts to the industrial centers, no serious hardships were expected.

In Occupied Zone.

Separatism.

The Separatists continued their activities in the occupied area during this quarter. A conference of delegates of the Rheinische Volksvereinigung (Rhenish People's League) was convened at Boppard on July 2nd and was attended by Dr. Dorten, who discussed the critical situation and advocated reorganization of the League on a federal basis.

The delegates of the Rhenish Republican People's Party held a meeting at Aix-La-Chapelle on July 23rd, about 600 persons being present. Herr Smeets, the re-elected Chairman of the Party, urged the foreign representatives present to work for the support of the movement abroad. He strongly advocated "the liberation of the Rhineland from the Prussian yoke, the expulsion of all Prussian officials and government employees and the election of Rhenish representatives to be attached to the Interallied Rhineland High Commission." Another Separatist group, the so-called Christian People's Party, came into the lime-light early in August. This party was a group of people who had separated from the Centrum Party two years before and formed two wings, one of which pursued Dr. Dorten's policy, while the other restricted itself to advocating radical federalism. This latter group was represented in the Reichstag by Dr. Deermann, of Cologne, the only deputy of the Party. He placed himself in opposition to the Dorten wing, with the result that this group, which in the meantime had amalgamated with the Rhenish People's Association, expelled him from the Party after repeated warnings to do so. Dr. Deermann, in a public statement, called this expulsion a "joke on Dorten", saying that he had never belonged to Dr. Dorten's particular wing of the so-called Christian People's Party of the Rhineland.

As the summer passed, it became apparent that the cause of the Separatists was more or less of a failure. In September, it was noticed that the movement, because of its pronounced failure in the cities, was being carried into the rural districts. Herr Smeets and his agents seemed to hope that the unsophisticated peasant population would furnish a more fertile field for their otherwise unpopular efforts ; but there, also, they made little or no headway.

Towards the end of September, another newspaper,

"Der Republikaner" (The Republican) was added to the list of publications issued at Coblenz. This new daily was promoted by the "Rhenish Republican People's Association", an organization of about fifty members, under the leadership of a city councillor in Bad Ems by name of Kaffine. The fact that the Coblenz Dorten concern printed the new paper left no doubt regarding its policy. In taking cognisance of the appearance of "Der Republikaner", the non-Separatist Rhenish press expressed the belief that the new organ was intended to spread the doctrine of Separatism among republicans of the Rhineland, in contradistinction to the older Separatist sheets, which appealed mainly to reactionaries.

In American Area.

Effects of Rathenau Murder.

The nation-wide agitation in favor of the republican form of government and the intense political strife which swept over Germany following the murder of Foreign Minister Rathenau, found its reflection in the Coblenz City Council early in July. The Majority Socialist group made an unsuccessful attempt to pass a bill calling for immediate purchase of a republican flag and renaming of a number of Coblenz streets which bore names closely connected with the old régime. As was expected, the motion found no favor with the Bourgeois groups. At its session of July 12th, the City Council, without debate and against the votes of the Socialists, rejected the proposals.

A mass meeting was held at Coblenz on July 4th in response to a proclamation of the Free Trade Unions and the

Socialists, urging the labor organizations to demonstrate in favor of the new bill for the protection of the Republic. The speaker of the day, a local leader, regretted the small attendance and ascribed it to counter-propaganda originating with the Communists, who had spread a rumor to the effect that the meeting had been prohibited by the American authorities.

Dr. Fuchs becomes Ober-Praesident.

In July, the Prussian Cabinet selected Dr. Fuchs, President of the Government District of Treves, to be the successor of Herr von Groote as Ober-Praesident of the Rhine Province. The new chief executive, though closely associated with the Centrum Party, had never played an important role in party politics. At the time of his appointment he was 47 years of age, having been in the government service since his graduation from the University of Bonn. He had been President of the Treves District since 1919.

Retirement of Monarchist Officials.

About this time the Prussian cabinet ordered the retirement from office of a number of presidents of Prussian government districts for general political reasons. Among these was Herr von Groening, President of the District of Coblenz. Herr Bauknecht, a member of the local group of the Majority Socialist Party, was named as the likely successor to Herr von Groening. While the appointment of Dr. Fuchs as Ober-Praesident met with general approval, the probable nomination of Herr Bauknecht resulted in a sharp controversy between the German People's Party and the Centrum, on one side, and the Majority Socialists on the other. However, the Rhenish Provincial Committee flatly refused to indorse the Prussian Cabinet's Socialist candidate, the Bourgeois vote being unanimous against his appointment. The Committee's approval being required under the constitution, it was not believed that Herr Baumecht would be able to secure the coveted appointment.

Party Politics.

Towards the end of the third quarter of 1922, with the reopening of the Reichstag approaching, party political activities revived throughout the American area. Practically all parties held meetings, at which prominent speakers discussed the important foreign and domestic political and economic problems. The two largest meetings were held by the Centrum and the German People's Party late in September and early in October, the G.P.P. being addressed by the national chairman, Herr Stresemann. The latter, in announcing a change of policy, stated that the G.P.P. was ready to cooperate with any party, including even the Socialists, in the reconstruction of Germany.

Economic Conditions.

The American area did not escape the consequences of the abnormal depreciation in value of German currency during the third quarter of 1922. Prices rose automatically as the mark declined. Shop-keepers, who, to prevent discrimination against the troops of occupation, were required by American orders to mark prices plainly on all goods, were kept busy changing the price-tags in an effort to maintain their receipts on a level enabling them to replenish their depleted stocks.

Because of the danger that the entire contents of the shops might be bought up by foreigners whose currency had such great purchasing power, the administrative authorities of Coblenz, in collaboration with tradespeople and consumers, put certain regulations into operation which restricted sales to all foreigners other than members of the Armies of Occupation or the Interallied Rhineland High Commission.

The following prices of various commodities, picked at random and shown for July 1st and October 1st, give an illustration of the trend of the cost of living:

(In Marks).

	July 1st, 1922.	October 1st, 1922.
Onions, per lb.	14.-	14.-
Potatoes, " "	3.30	6.50
Carrots, " "	10.-	6.-
Margarine," "	41.-	220.-
Butter, " "	85.-	420.-
Eggs, each	6.80	26.-
Beef, " "	60.-	170.-
Veal, " "	56.-	180.-
Pork, " "	73.-	260.-
Chicken " "	70.-	200.-

The following table contains a further comparison of prices in the American area with the space of a few weeks:

(In marks).

	July 19th.	August 23rd.	Sept. 12th
Men's shoe repairs (soles and heels)	245.-	650.-	800.-
Women's " " "	220.-	550.-	700.-
Children's " "	168.-	400.-	550.-
Soap, per cake	59.-	120.-	200.-
Soda	8.-	14.-	26.-
Soap Powder	40.-	72.-	110.-

The steady increase in the cost of living naturally

led to demands for higher wages on the part of working-people and employees. These were always granted either voluntarily or after arbitration had been resorted to, so that no serious labor disputes were experienced. At the end of the period, the following wages were being paid at Coblenz :

Bricklayers	74.90 marks	per	hour.	
Bricklayers' helpers	71.20	"	"	"
Carpenters, construction work	75.40	"	"	"
Carpenters, other work	74.-	"	"	"
Excavators	70.90	"	"	"
Metal workers, including plumbers, Blacksmiths, locksmiths, etc.	7?.-	"	"	"
Roofers	75.-	"	"	"
Tailors	66.-	"	"	"
Brewery workers	2,200.-		"	week.
	(increase to 4,000.0 marks pending).			

Bendorf Gas Works Strike.

A strike at the Bendorf Gas Works occurred in July. The strikers having exceeded the limits of reason, the Commanding General ordered two of his civilian officials, backed up by one company of Infantry, to settle the matter on the 15th. After the arrest of two or three of the leaders, the affair ended.

Celebration of Holidays.

The 4th of July was fittingly celebrated by the American Forces in Germany wherever troops were stationed in the American area. A holiday was declared and sports of all sorts were entered into between Americans and Allies. A magnificent display of fireworks on Ehrenbreitstein lured the American colony and thousands of Germans to the Rhine at night. All the Commission heads called on the Commanding General in turn in honour of our national independence and telegrams were received from all the army chiefs.

On July 14th, the first Bastille day since the arrival of the French troops in the Coblenz bridgehead, General Allen and Colonel Piazza, the senior French officer in the American area, reviewed and inspected the French troops under their command. The remainder of the day was given over to sports of various kinds. The Commanding General sent a letter to M. Tirard expressing esteem and friendship for France and a telegram to General Degoutte in the same vein.

July 21st, the Belgian national holiday, was also appropriately observed in the American area.

Colonel Piazza, the senior French officer present in the American zone, was given permission to celebrate Chateau Thierry day - July 25th - by a mass on Clemensplatz, games at Carnival Island and a dance at the club. A request to have a big torch light procession through the streets with drums and trumpets, the night before, was refused, because it appeared too much like a part of a studied program to emphasize French holdings and activities in the American area.

Colonel W. H. Harts C. of S.

Colonel W. H. Harts, who was Chief of Staff, A.F.G., in 1919-20, arrived at Coblenz on leave on July 24th. A cable was sent to the War Department on the 25th, requesting authority to put him on duty with the A.F.G. On July 31st, the War Department directed him to report to the Commanding General for duty. He assumed the duties of Chief of Staff on August 1st, relieving Lieutenant Colonel H.H. Sheen, who had been the Acting Chief of Staff since the departure of Brigadier General W. H. Hay.

Allied Issues.

General the Earl of Cavan, Imperian Chief of Staff, British Army, and his deputy chief of staff, Lord Chetwode, arrived at Coblenz to confer with Lord Kilmarnock on August 23rd. General Allen had a conference with them, at which it developed that Marshal Foch had refused to give Bonn back to the British as requested ; that England opposed in principle any independent action by France against Germany, but would do no more than express that thought ; and that since prohibition of requisitions in the British zone in case of such independent action would be of no practical value, no protest would be made.

The question of a French advance into the Ruhr, or sanctions, or both, assumed serious proportions in August, After a conference with Colonel Stone and Lord Kilmarnock on the relation of such a move to zones other than the French and the lines of action open to the American and British representatives in such an event, General Allen dispatched a detailed cable to the State Department giving a complete picture of the situation in the Rhineland.

A letter from an official in the State Department received by General Allen in September, indicated that should France go into the Ruhr, it would be better for the A.F.G. to be withdrawn at once, without any exchange of letters, thus letting France know in unequivocal terms that its policy was disapproved.

Franco-American Issues.

General Degoutte called on General Allen about this time to talk over military matters. He proposed sending a battalion supported by cavalry and artillery to each of the towns of Montabaur and Dierdorf. General Allen countered by suggesting that one of the battalions then near Coblenz be sent to Montabaur, especially since there were too many French troops in the streets of Coblenz and American representation would suffer if such a condition were allowed to continue.

The Allied Commander then requested General Allen to change his orders so as to permit French troops in the American zone to carry bayonets when off duty. He contended that the absence of bayonets was a strong incentive for the French soldiers to carry concealed weapons. General Allen stood fast on treating the French the same as the Americans and could see no reason why the French would be less aggressive or get along with the Germans any better by being armed at all times.

The next request was to have a section of French gendarmerie sent to this zone to work with the American military police and the Germans. General Allen also refused this request and said that he would call on French N.C.O's when he needed them.

General Allen impressed the Interallied Commander in Chief with the fact that it would be better to diminish the French representation at Coblenz than increase it.

All differences were adjusted in an amicable way.

The Comptroller General of the French Army requested permission on August 11th to establish a small bureau of military administration to facilitate the supply and requisitions required by the French troops. He was informed that requisitions on the Germans must remain in General Allen's hands ; that the method in force was satisfactory and had brought forth no complaints from the French troops in this zone and that the American government was properly sensitive in regard to the American status in this zone.

It was learned in September that additional French troops were being brought into the French zone quietly and that school facilities were being sought for 150 children of the two and a half French regiments in the American zone, as against facilities for fourteen American Army children.

Another attempt at encroachment emanating from French G.H.Q. determined the Commanding General to establish the American position firmly in the minds of all concerned during the early stage of the dual command. He consequently prepared a communication for General Degoutte and furnished a copy to M. Tirard. This communication established certitude at the Interallied High Command and at the High Commission that the American zone remained American and was commanded by General Allen.

M. Tirard again declined to have any committee or other agency investigate the expenses of the High Commission. The other Commissioners consented to such an investigation as requested by the Army Costs Conference.

Manoeuvres, Reviews and Inspections.

The Commanding General, accompanied by the Chief of Staff, attended the final day of the French manoeuvres near Euskirchen on September 22nd. He also attended the first day of the manoeuvres of the French 33rd Corps, under General Jacquemont, which involved the crossing of the Rhine near Bonn. The manoeuvres lasted several days.

The inspection and review by General Allen of parts of three French regiments constituted a unique incident in the American occupation.

The Commanding General was present at the 8th Infantry manoeuvres near Bassenheim on September 28th, when the regiment made its usual good showing.

Visits of Marshals Fayolle and Pétain.

Marshal Fayolle of France paid a visit to Coblenz during this quarter and was received at Headquarters by General Allen.

Marshal Pétain of France was another visitor.

Fourth Quarter of 1922.

Economic Crisis and Reparations.

The rapid decline of German currency towards the end of the year 1922 and the economic crisis caused thereby not only governed domestic politics but also had the effect of bringing the reparations problem and the compensation adjustment negotiations into the foreground once more. The German reparations problem was given sudden prominence when Mr. Reginald McKenna, the noted British financier and former Chancellor of the Exchequer, addressed a conference of American bankers at New York on that subject. After dwelling on the low rate of German exchange and its consequences, he expressed the opinion that although Germany was certainly in a position to pay "something", yet its payments could be made neither in the form nor under the conditions that had been imposed upon it. Sir John Bradbury, British representative on the Reparations Commission, submitted a plan to that body in October for the solution of the difficult question which took Germahy's economic distress into consideration. The French member of the Commission, too, advanced a new plan, having in view a closer supervision of German finances in order to stabilize the mark, without which action he considered it impossible for Germany to remain solvent. The members of the Commission then decided to go to Berlin for the purpose of studying the problem on the spot. They arrived at the German capital on October 31st. The discussions there lasted until November 11th, when the Commission returned to Paris.

German Proposals to Reparations Commission.

The German proposals submitted in the course of the negotiations proved to be unsatisfactory. After the Commission's departure from Berlin, the result of the discussion there was summarized in a German note dispatched to the Commission on November 13th. This note contained the following provisions:

1. In order to stabilize the mark, a definite settlement of the reparations question, in accordance with Germany's capacity to pay, is necessary ;

2. A moratorium must be granted on all payments in cash and kind for a period of three or four years ;

3. An international credit, supported by the Reichsbank to the extent of 500,000,000 gold marks, yielding 1,000,000,000 gold marks in all, shall be raised for the purpose of improving the mark ; Germany will then take all proper steps towards balancing its budget and increasing its production ;

4. Independence of German economic administration must be restored in the occupied area.

Resignation of Cabinet.

Cancellor Wirth obtained the approval of all the parties, the German Nationals and Communists excepted, before the note was submitted. Although he thus succeeded in rallying all the moderate parties of the Reichstag to subscribe to his policy as outlined in it, he nevertheless failed to bring about unanimity between the United Socialists and the three Bourgeois parties (G.P.P., Centrum and Democrats) on the question of the much discussed big coalition extending from the Socialists to the German People's Party, inclusive. In order to settle this matter, he called upon the Socialists to make a definite decision on the issue, whereupon they declared that they would not join a coalition in which the G.P.P. was represented. This resolution prompted Dr. Wirth to present the resignation of the entire cabinet to President Ebert on November 14th. The President immediately accepted it, with the proviso that the old cabinet should continue in office until a new one had been formed.

Dr. Cuno Forms Cabinet.

After lengthy deliberations with the leaders of the political parties, the National President entrusted Dr. Cuno, General Manager of the Hamburg-American Line, with the task of forming a new cabinet on November 16th. Dr. Cuno accepted and at once set to work to carry out his mission. In view of the fact that the United Socialist Party again refused to participate in any cabinet in which members of the German People's Party were represented, the new one was formed of Bourgeois representatives and experts only, several of the members of the Wirth cabinet having been retained.

The new cabinet, which was described by Dr. Cuno as a "cabinet of work", was presented to the Reichstag on November 24th. A list of the ministers follows :

Chancellor : Dr. Cuno (no party; favored G.P.P.);
Minister of the Interior : Oeser (Democrat);
Minister of Foreign Affairs ; Dr. von Rosenberg (no party);
Minister of Justice ; Dr. Heinze (G.P.P.);
Minister of Economics : Dr. Becker, (G.P.P.);
Food Minister : Dr. Mueller, Bonn (Centrum), later
Dr. Luther (no party; favored G.P.P.);
Minister of Labor: Dr. Brauns (Centrum); retained portfolio;
Minister of Finance : Dr. Hermes (Centrum) retained portfolio;
Minister of Treasury : Dr. Albert (no party);
Minister of National Defence : Dr. Gessler, (Democrat);
retained portfolio ;

Minister of Transportation : Groener (favored Democrats); retained portfolio);
Postmaster General : Stingl (no party, favored Bavarian People's Party).

Dr. Mueller, the new Food Minister, submitted his resignation on the same day an account of the relations with the Rhenish Separatists which he had displayed in the Rhineland during 1919.

Chancellor Cuno delivered a very forceful speech before the Reichstag in setting forth his policy on November 24th. Basing his program on the German note of November 13th to the Reparations Commission, he proposed an increase of production by a proper regulation of working-hours, without, however, abolishing the eight-hour day. He touched upon the Rhenish question and declared that the government would never consent to a surrender of German soil. He counselled unity and exhorted the people to utmost effort in view of Germany's precarious situation. The Reichstag took cognizance of the Chancellor's statements and, against the votes of the Communists, the Ledebour Party and the three adherents of the "Deutsch-Voelkisch" (Anti-Semistic) Group, approved the government's policy as based on the note of November 13th.

Agreement on Compensation Payments.

The German government's negotiations with the representatives of the foreign compensation offices led to an agreement in October, according to which Germany was to make no more payments until July, 1923. This decision was hailed with general satisfaction in German official circles, which declared that it established an admittance of Germany's inability to render cash payments for some time to come.

During the same month, Chancellor Wirth made several speeches in which he attempted to establish "the fallacy of Germany's sole responsibility for the war." He had the nation's full support in his efforts to prepare the world for a revision of the Versailles treaty, although the parties of the Right considered the raising of that issue their special privilege.

German Government Makes New Proposals.

The Prime Ministers of Great Britain, France, Italy and Belgium met at London on December 9th to engage in preliminary discussions regarding the reparations problem and the adjustment of interallied debts in connection therewith. In a letter addressed to the British Prime Minister and delivered to Mr. Bonar Law by State Secretary Bergmann, the German government submitted new proposals for the solution of the reparations question, which may be summarized as follows:

> "The German government undertakes to stabilize the mark with its own resources if necessary, but asks that its reparations obligations for the next few years be regulated in conformity with its note of November 13th and that it be granted equality of rights in commercial intercourse with foreign nations. It proposes to float loans in Germany and abroad, the proceeds to be utilized for reparations purposes, and it is prepared to enter into immediate negotiations with the Allies for the purpose of coming to an agreement on this basis."

Rejection of German Proposals.

The German proposals were rejected by the Allied Prime Ministers on December 11th, on the grounds that the plan was unsatisfactory. In view of the fact that, in the time available, it had not been possible for the Allied Prime Ministers to arrive at definite conclusions on the issues involved, they decided to adjourn the London discussions and to continue the conversations at Paris on January 2nd. In case the preliminary deliberations scheduled for Paris should reach a successful conclusion, a general conference of all the interested governments was to be called immediately thereafter.

Default by Germany.

The Reparations Commission established a default on the part of Germany on December 26th on account of its failure to deliver the stipulated quantities of timber to the Allied Powers. The Commission further decided that this failure constituted a default in the sense of Paragraph 17, Annex II, Part VIII, of the Versailles treaty, and that the attention of the Allied governments should be invited to the reparations note of March 21st, according to which Germany must make good all shortages by additional cash payments. The German government had previously submitted a statement giving the reason for the delay in the timber deliveries and asking for an extension of the period of delivery until April 1st, 1923.

After thorough deliberations with representatives of industry, banking, commerce, trade and labor, the German government unanimously decided to submit new reparations proposals to the Paris conference of Allied Premiers. This offer, according to the German press, would be "the last effort of an exhausted people."

Chancellor Cuno defined the government's attitude on the repartions question before a meeting of the Hamburg Commercial Society on December 21st. He deprecated the policy of pledges and endeavored to prove that French apprehensions in regard to a German aggression were unfounded. He declared that the final settlement of the question must do away with all obligations other than those of reparations derived from the Versailles treaty. He demanded, that economic and political freedom and equality among nations be restored to Germany and that the foreign troops gradually evacuate the occupied zones.

Rumors of American intervention.

The London conversations were scarcely terminated when an announcement appeared in the press to the effect that the United States might give up its policy of isolation and participate in the solution of the momentous problem that was occupying the minds of European statesmen. The exaggerated rumors of a forthcoming American loan in favor of German and the belief that the United States would once more play an active part in European politics affected the money market and caused a temporary improvement of the mark at once. However, the majority of the German press came to the conclusion that while the United States government was undoubtedly willing to cooperate in the settlement of the reparations question, yet nothing had occurred to justify over-optimistic expectations. This opinion prevailed, in spite of the fact that Secretary of State Hughes, in a speech at New Haven, Conn., on December 29th had suggested calling a commission of economic experts, rather than of statesmen, to fix the definite amount to be paid by Germany in reparations, apart from political considerations.

Incidents at Stettin, Passau and Ingelstadt.

The tension of the foreign situation was accentuated by a note from the Ambassadors' Conference to the German government on December 1st. This note referred to incidents at Stettin, Passau and Ingelstadt, where mobs had made attacks upon officers belonging to the Interallied Military Control Commission. The incidents at Passau on October 24th and at Ingelstadt on November 23rd were considered particularly serious. Its demands for an apology and a compensation not having been satisfactorily complied with by the German government, the Ambassaors' Conference stated that a fine of 500,000 gold marks should be paid by each of the two towns before December 10th. It further demanded that the Prime Minister of Bavaria should apologize in writing to the Control Commission for the incidents, and declared that if the payments were not made or were made only partially by December 10th, the Allied governments would seize the revenues accruing to the Bavarian government from the Palatinate.

On December 10th, the German government complied under protest with the demand for payment of 1,000,000 gold marks, stating that the National government would place that amount voluntarily at the disposal of the Interallied Military Control Commission. The demand for a special apology on the part of the Bavarian Premier was not complied with, on the grounds that the national government alone was entitled to represent Germany and the federal states abroad.

The German manner of compliance caused the dispatch of another Allied note, which was delivered on December 19th and which provided for imposition of certain local sanctions on the cities of Stettin, Ingelstadt and Passau. These pertained chiefly to the dismissal of the responsible police officials and military commanders.

Question of Replacing Interallied Military Control Commission.

The question of replacing the Interallied Military Control Commission in Germany by a military guarantee committee was taken up by the Ambassadors' Conference towards the end of the year. After some deliberations, however, that body informed the German government that certain stipulations, pertaining especially to reorganisation of the police forces and so-called transformation of plants manufacturing war material, must be carried out before a termination of the activities of the Interallied Military Control Commission could be considered.

Party Politics.

The National Democratic Convention was opened at Elberfeld on October 8th with a demonstration in favor of the Rhineland. The Interallied Rhineland High Commission was made the subject of criticism and the necessity of a strong centralised state to avert disintegration of the nation was emphasized. The convention closed on October 10th. The advisability of forming a cooperative working union of the Bourgeois parties was debated, but no definite conclusion was reached.

A small group of malcontents, under the leadership

of Herr Ledebour and made up of former adherents of the independent Socialists, refused to join the United Socialist Party and adopted the designation of "Ledebour Party". Another group of three members of the Reichstag seceded from the German National People's Party and formed the "Deutsch-Voelkisch" Group. The platform of this new party was based upon ultra-radical, monarchistic and anti-Semitic principles.

Elections.

The elections to the Saxon State Assembly, which were held on November 5th, resulted in a victory of the Communist Left. The efforts of the Bourgeois parties to break up this majority and overthrow the purely Socialist government proved to be unsuccessful. The members of the old Socialist government of Saxony were reelected by the State Assembly on December 5th, the Communist deputies voting for the government.

Elections to the Reichstag, The Prussian State Assembly and the Upper Silesian Provincial Diet in the German portion of Upper Silesia were held on November 19th. The outstanding feature of these elections was the fact that the Poles did not succeed in obtaining a single seat in the Reichstag and that they obtained only one in the Prussian Assembly. The Reichstag elections gave the Centrum Party three seats, and the German Nationals and the United Socialists one seat each. In consequence of the Upper Silesian elections, the number of Reichstag deputies was reduced from 469 to 459.

Municipal and communal elections held in November in various States showed considerable losses of the Democrats and the United Socialists in favor of the parties of the Right.

Reichstag.

On October 24th, against the vote of the Socialists, the Reicstag passed the bill providing for an increase in the price of the first third of grain to be delivered under the government grain control act. It also extended President Ebert's term of office until June 30th, 1925.

The nation-wide demonstrations for the Rhineland as a result of reports of alleged French aspirations concerning it and the Ruhr District were concluded by a big meeting in the Reichstag on December 7th. A resolution was passed stating that in "this fateful hour", the inhabitants of unoccupied Germany repeated their former solemn promise to remain loyal to the harassed Rhineland, and that the recent threat against the Rhineland had found a united Germany ready to defend it.

Bavaria.

The political confusion throughout Germany in the month of October was greatly intensified by increased activity of the Bavarian monarchists and by the extreme federalism expressed in the Bavarian People's Party's new platform, as well as the general particularist attitude of the Bavarian government. This continous agitation brought about the downfall of Bavarian Prime Minister Count Lerchenfeld on November 2nd. Count Lerchenfeld decided to resign after being convinced that he no longer enjoyed the confidence of his party (Bavarian People's Party), in which the reactionary, particularist and fanatical elements had gained the upperhand. The Bavarian

State Assembly elected Herr von Knilling, former Minister of Public Worship and Education, to the post of Prime Minister on November 8th. Herr von Knilling stated in a programmatic speech that he would continue the policy of the coalition parties, all the essential points of which were still applicable. He disclaimed any connection with separatist movements in Bavaria, criticised Chancellor Wirth's "policy of fulfilment" and voiced the hope that the national government would take up the fight against the "myth" of Germany's sole responsibility for the war with greater energy than before.

Notwithstanding Minister von Knilling's serious warnings against disturbances and social upheavals, the influence of the National Socialist Party increased considerably throughout Bavaria. This party had a decided anti-Semitic character and aspired to national dictatorship. It was headed by Herr Hitler, a German-Austrian who had become extremely popular in Bavaria. Indeed, the National Socialist Party became all-powerful in Bavaria. In imitation of the Fascisti movement in Italy, their "shock-troops" even invated Wuertemberg, where serious encounters occurred between them and the local workers of several cities, resulting in casualties on both sides.

Hanover and East Prussian Separatists.

The National Minister of the Interior vetoed a motion of the German-Hanoverian Party (Guelphs) demanding a referendum in the districts of Lugneburg and Stade for the purpose of deciding the question of separation from Prussia. This decision, however, did not prevent the Guelphs from forwarding another motion to the national government with the same object. Even in East Prussia, a movement was started with the object of converting that province into an independent federal state within the German nation, However, the objection to the creation of an East-Prussian federal state expressed by the leaders of the national political parties doomed the aspirations of the East-Prussian separatists at the very beginning of the movement.

Rate of Exchange.

General economic conditions at the beginning of the last quarter of the year seemed to promise a period of comparative calm and an adjustment of prices, wages and the cost of living to the new level of the rate of exchange. However, a new slump in the exchange rate beginning on October 10th upset all expectations, for it soon caused the mark to be quoted at less than 3,000 to the dollar, i.e., a depreciation of more than 90%. Since this decline could hardly be laid to the reparations payments, except in as far as they perhaps influenced the already abnormal inflation, it was apparently caused primarily by unfavorable domestic conditions.

One of the most detrimental factors affecting the situation was the unprecedented activity of the Berlin banknote press which, in an effort to supply banks and business establishments with the enormous amount of currency they required because of the monetary depreciation, printed no less than 10,000,000,000 marks a day. Even this tremendous output, however, was not sufficient to supply the ever-increasing demands. Mr. Lloyd George's resignation was apparently another cause of the financial crisis, many Germans having looked upon the famous Welshman as a strong influence in Germany's favor. His retirement gave a new impetus to the downward rush of the mark. During the last days of October, it was quoted at about 4,500 to the dollar. This movement continued during the first days of November at a rate exceeding anything thitherto known and threatened to eclipse even the record established by the Austrian crown in the past. In one day, for instance, the dollar quotation in marks jumped from 6,900 to about 8,900, while the average decrease during the first part of the month was about 100%.

After Dr. Wirth's resignation and the appointment of Dr. Cuno as Chancellor, the mark recovered more than 2,000 points and remained comparatively steady until the middle of December. This was due in no small measure to political developments in Germany, especially the appointment as Chancellor of a leading businessman and the formation of a cabinet consisting largely of experts. In spite of this improvement, however, there was no change in the general economic situation and German financial experts agreed more than ever that a stabilization of the mark and an appreciable improvement in the country's economic organization could not be brought about until Germany had been granted a due respite of payments arising from the Versailles treaty and unless the budget be made to balance by the requisite augmentation of internal revenues. The mark suddenly improved considerably in the middle of December, unofficial quotations even reaching as low as 5,400 at one time. Later, however, the dollar regained its strength and was quoted in the neighborhood of 7,000 at the end of the year.

The principal cause of the improvement of the mark was the alleged change of attitude of the United States government towards Europe. The many rumors and reports emanating from Washington led to the belief that the oft-predicted "American intervention" was at last about to take place. The hope was expressed throughout Germany that this intervention would operate to remove the causes of the depreciation of German currency and of the generally unfavorable situation. That this view was shared in the United States was shown by the fact that the mark rose even more rapidly on the New York Stock Exchange than in Europe. Rumors of a large international loan to be floated in favor of Germany, which, it was variously stated, would total from $1,500,000,000 to $2,000,000,000, added to the general optimism.

Floating Debt.

The German floating debt increased by leaps and bounds during the first ten-day period of December. Receipts from all sources, including the compulsory loan, amounted to only one seventh of the sum expended. In order to meet the remaining six-sevenths of expenditures, the floating debt had to be increased by no less than 123,190,120,000 marks, bringing the total debt up to 1,353,588,000,000 marks, or double the October amount and four times the total announced at the end of

July. The record deficit for this period was officially attributed by the government to the usual expenditures on account of the Versailles treaty, the necessary purchases of grain from abroad and the unusually large sums needed by the national railways.

Governmental Departments.

All the efforts of the heads of the various governmental departments to economize and reduce expenditures were unsuccessful, with the exception of the Post Office Department, which continued to show considerable earnings and gradually recovered the big deficit accumulated furing the early part of 1922. The Railway Department continued to be a heavy burden of expense to the government, the last ten-day period of December, for instance, showing a deficit of no less than 49,130,000,000 marks. This big deficit was explained by the large advance purchases of necessary raw material that had been made. The National Ministry of Transportation raised railway fares and freight rates by 100% on November 1st, but even this increase failed by far to bring the prices up to a level reasonably proportionate to thelow value of currency. The postal rates were also raised by 100% on December 15th.

Cost of Living.

The cost of living continued to rise concurrently with the decline of the mark. Prices for all commodities changed from day to day, so that the poorer, and especially the middle, classes, encountered great difficulties in purchasing supplies for their daily or weekly needs. In spite of this, the employment situation remained favorable, as employers always agreed on higher wages to meet the changing conditions.

The panic-like demand for goods and foreign currencies early in November drove prices to a height out of all proportion to the increase in the circulation of currency. Bank exchanges for loans were driven up to a minimum of 20% on government bonds and as high as 50% and more on less acceptable securities. The later improvement in the value of the mark did much to put an end to these abnormal business activities.

Food Situation.

Semi-official reports declared that the grain supply on hand in Germany would last until February, 1923, and that an enormous amount of paper marks would have to be spent for purchase abroad of a sufficient quantity of grain to tide the country over until the next harvest. A tremendous increase in the price of bread was consequently inevitable. As more or less serious food riots involving much plundering and some bloodshed occurred late in October and in the early part of November in several big cities, including Berlin, Hamburg, Cologne and Dresden, the authorities adopted precautionary measures in order to cope with the serious situation.

The index figure for wholesale prices dropped by 8.4% early in December as a result of the comparatively stable rate of exchange. This was immediately made the subject of much optimistic comment by a large section of the press, which welcomed this unexpected decline of prices as an indication of a greater improvement soon to follow. Less optimistic circles, however, realized that it was due almost entirely to the sudden rise and stabilization of the mark, by which prices for such commodities as had to be purchased abroad were chiefly affected. No improvement of the situation on the retail market was observed.

According to official figures, the potato crop was estimated at the enormous total of 39,240,978 tons. It was stated that this amount would amply suffice during the winter months.

Living expenses did not show such a tremendous increase in the month of December as they did in November, because of the comparative steadiness of the exchange rate during the last month of the year. There was, however, much distress ih Germany, the papers were full of appeals for aid to the poor and people were urged to do everything possible to relieve "the great suffering."

<u>Fuel.</u>

The question of reducing the compulsory coal deliveries to France occupied much space in the economic section of the press toward the end of the year. It was estimated that the monthly cost of producing this coal was about 20,000,000 marks, by which amount the paper money put in circulation by the Reichsbank was consequently increased every month, there being no receipts to cover the expenditure. It was also stated that German industry was paying at least

40,000,000,000 marks for British coal to replace the coal thus going to France and Belgium. Although the overtime agreement in the Ruhr District led to an increase in output, this did not influence the generally unfavorable situation, which was one more reason why the German press expressed the belief that a stabilization of the mark was an impossibility. The fuel situation remained unsatisfactory until the end of the year, due in no small measure to the fact that about 1,000,000 tons of coal had to be imported from England each month.

Shipping and Foreign Trade.

Reports from Hamburg, Bremen and other German ports indicated continued activity in shipping, overseas trading and shipbuilding. The reestablishment of an old line or the opening of a new one was announced by one of the large steamship companies almost weekly during the month of October, and the launching of new vessels was the order of the day. It became evident that the German merchant marine was well along on the road to normal conditions.

Latest statistics available on Germany's foreign trade showed that the excess of imports over exports was still growing. According to the report for September, for instance, imports for that month amounted to 421,300,000 gold marks, against 280,400,000 gold marks of exports. The press carried many reports indicating that Germany's trade with Russia was reviving as rapidly as circumstances permitted. As a matter of fact, at least twenty-five important trading agreements were concluded between German industrials and the Soviets since the Rapallo treaty had come into existence.

Labor.

There was little change in the labor situation, the usual number of strikes being reported, although none of a serious nature or long duration with the exception of the Ludwigshafen chemical workers' strike. This lasted about six weeks in the months of November and December and was instigated by the Communists for more or less political reasons. In their usual demands for higher wages, the workmen's and employees' organizations met with determined opposition on the part of the government during the last days of December. Although the railwaymen, especially, insisted upon more pay, a repetition of the last year's railway strike, which was predicted, did not take place since, in view of the complicated foreign situation, all parties, the Communists excepted, cooperated with the government and induced the various trade unions to suspend their demands.

In view of the fact that the United Socialist Party passed a resolution at this time calling for the preservation of the eight-hour day under all circumstances, the belief was generally expressed that an increase in production would have to be made by some method other than that of increasing the number of working-hours.

Success of Cuno Government.

In general, Germany's economic situation at the end of the year 1922 appeared to be improving somewhat. This turn of affairs was primarily ascribed to the Cuno government, which was also credited with the comparative stability and improvement of the mark.

In Occupied Zone.

Smeets Case.

After thorough examination of the documents pertaining to the Smeets case, the Interallied Rhineland High Commission early in October stayed the execution of the sentences imposed upon the Separatist leader by the courts of Cologne. This action was later the subject of debates in both the Reichstag and the Prussian Assembly, in which it was attacked as an unlawful interference in the internal affairs of the nation. An attempt was made to conceal the real feeling it excited, but it was nevertheless observable that actual hostility had been aroused against the occupation of the Rhineland in general and against the Commission in particular. The press was highly indignant and declared that relations between Rhinelanders and the Commission would suffer in consequence. The German government lodged a protest with the governments represented on the Commission and requested that the members of that body be instructed to reconsider and revoke their decision at an early date. The "Prussian Association of Judges", at a meeting held at Berlin, passed a resolution protesting against the decision, which it called "an interference in Germany's sovereign rights."

Separatism.

The Separatists displayed increased activity in the Rhineland in November, making the most of the economic crisis for furthering their cause. While the members of the Dorten faction apparently contented themselves with insisting upon an immediate referendum which they hoped would lead to statehood for the Rhineland, the Smeets group came out with a flamboyant manifesto urging the Rhenish population to separate from "Prussianized Germany" and to form a republic with a distinct gold standard and an orientation westward.

Reports concerning certain French intentions of confiscating the Rhineland and occupying the Ruhr district came from Paris in December. These naturally stimulated the activities of the Rhenish Separatists. The Smeets group seemed to be expanding. A meeting held at Treves, which no outsider was allowed to attend, was addressed by the over-active Smeets on the tasks of Separatists in the near future.

Herr Smeets continued his propaganda with great vigor as the year neared its end. He held an "extraordinary" congress of native Rhinelanders at Bonn on December 17th. Notwithstanding the fact that strict control was exercised at the meeting, certain non-Separatist papers were able to report lengthily on the discussions. According to them, Herr Smeets declared that "the great goal would be attained within a very short time, although he was unable to name the exact date on which the seizure of the Rhineland by true Rhinelanders would be accomplished."

Despite all the propaganda and agitation with which the Separatists inundated the Rhineland during 1922, they were apparently unable to make any appreciable gains. By far the overwhelming majority of the Rhenish population remained loyal to Prussia and Germany, and its opposition to the Separatist ideas was manifested in hundreds of meetings and resolutions. Nevertheless, the Separatist cause seemed to grow stronger as distress in the Rhineland increased, and the danger of its severance from Germany, either by force or otherwise, took on more serious aspects.

In American Area.

Instalment of Ober-President Dr. Fuchs.

For the first time since the war, a Prussian Minister was officially present at Coblenz when Herr Severing, Prussian Minister of the Interior, installed Dr. Fuchs in his office of Ober-President of the Rhine Province on October 14th. Officials of the Ober-Praesident and representatives of the presidents of the government districts of Cologne, Aix-La-Chapelle, Coblenz and Treves, as well as Prince Hatsfeldt, National and State Commissioner for the Occupied Territory, attended the ceremony. Speeches were delivered by Herr Severing and Dr. Fuchs, the loyalty of the Rhineland to Prussia and the nation being the keynote thereof.

Herr Severing later attended a reception in the Coblenz City Hall, where Chief Burgomaster Russell reported on the city's precarious financial situation, the heavy burdens of the occupation and the great lack of housing.

Party Politics.

Former Prussian Premier Stegerwald, leader of the Christian Trade Unions, delivered a long speech on the situation in Germany at a Centrum Party meeting at Coblenz on November 4th.

"Revolution Day", the anniversary of the revolution

of 1918, was observed by the local Socialist group on November 9th. No suspension of work occurred during the day, since the Free Trade Unions and the Socialists had revoked their original decision calling for "no work" and for demonstrations by the labor classes.

The Communist Party of Coblenz, which previously had never comprised more than about twenty to thirty members and had decreased to still fewer at times, seemed to take on a new lease of life in the last quarter of 1922. It held meetings regularly and the active membership list contained more than seventy-five names by the end of the year. This sudden increase was the result of greater activity on the part of the Communists, aided by the prevailing abnormal economic conditions which seemed to make the poorer classes susceptible to their intensive propaganda.

Although the protest movement against the separation of the Rhineland from Prussia and Germany had gradually died down, two more Democratic and Socialist demonstrations in favor of a "German" Rhineland were held in the American area in December. Both meetings passed the usual resolutions, the Democrats especially protesting against the alleged French plans concerning the Rhineland and the Ruhr District, which they declared incompatible with the peace treaty, true democracy and the right of self-determination of peoples.

<u>Dr. Brandt's Appointment as President of Coblenz Government District.</u>

The Provincial Committee approved the appointment of Councillor Dr. Brandt of the Coblenz Ober-Praesidium as

President of the Coblenz Government District late in December. Dr. Brandt who, prior to the appointment of Dr. Fuchs, was the acting Ober-Praesident of the Rhine Province, was an experienced official of old standing. He was born at Cologne and was 53 years of age in 1922. After having been Landrat at Simmern (Hunsrueck), for many years, he became Landrat of the Rural Kreis of Essen in 1914 and was appointed Ober-Praesidialrst (Councillor to Ober-Praesident) of Coblenz in 1918. Although not officially affiliated with any party, it was believed that he favored the G.P.P. The Coblenz Centrum press, which had fought so strenuously against the nomination of Herr Banknecht, the Socialist candidate, made no objection to Dr. Brandt's appointment.

Unsettled State of Labor.

The economic situation in the American area, as that in the remainder of Germany, was very precarious during the last quarter of 1922 as a result of the continued and rapid decline of the exchange rate. While there were no disturbances generally, despite the enormous increase in the cost of living and continual agitation on the part of Communists, yet the government's dilatory attitude in the matter of increasing wages and allowances of governmental officials and employees created discontent among these classes at times. The Coblenz branch of the Socialist Union of German Railwaymen participated therein and contemplated the idea of a railway strike in the occupied territory. Herr Kalt, chairman of the Coblenz branch, stated in a telegram to the Berlin authorities that unrest among the railwaymen was growing and that they were determined to act "radically". Although a local railway strike was proclaimed in the Palatinate early in November, it was speedily settled by the intervention of the occupying authorities there. The danger of an outbreak in the American area soon subsided. Apparently the main reason therefor was a lack of funds, because of which the labor organizations were reluctant about ordering a walk-out at the beginning of the winter.

Prices of Food.

The following prices charged for various commodities in the American area during the last quarter of 1922 serve to illustrate the increase in the cost of living :

(In marks).

	October 1st, 1922.	December 15,1922.
Onions, per lb.	14.-	23.-
Potatoes, per lb.	6.50	9.-
Carrots, per lb.	6.-	25.-
Cabbage (white), per lb.	6.50	35.-
Margarine, per lb.	220.-	1,100.-
Butter, per lb.	420.-	2,000.-
Eggs, each	26.-	90.-
Beef, per lb.	170.-	600.-
Veal, per lb.	180.-	600.-
Pork, per lb.	260.-	900.-
Chicken, per lb.	200.-	850.-

General Marty takes Command of French Troops in American Area.

General Marty, French Army, took command of the French troops in the American area on October 10th.

Visits of Prominent Germans.

Dr. Gustav Stresemann, President of the German People's Party, called on General Allen on October 9th and talked of Germany's domestic and foreign situation. German Minister of the Interior Severing, accompanied by Prince Hatzfeldt, called on October 14th. He stated to General Allen that all of

Germany was highly appreciative of the American policy in the Rhineland.

General Allen attends a Conference at Berlin.

General Allen attended a conference of American diplomats accredited to western European capitals at Berlin in October, Ambassador Houghton presided and the meetings were attended by Mr. Castle, in charge of the Western European Section of the State Department, Ministers Gibson (Poland), Washburn (Austria), Grew (Switzerland) and Brentano(Hungary).

This visit served not only to bring General Allen into touch with American representatives in other European countries, but afforded him an excellent opportunity to talk with various foreign officials and diplomats on questions bearing on America's participation in European affairs and to obtain their views of the Rhineland situation. He also had a long talk with Dr. Wirth on the general situation in Germany.

Commanding General at Brussels.

The Commanding General arrived at Brussels on November 10th in order to bestow the Congressional Medal on the Belgian unknown soldier in conformity with instructions from the War Department.

The ceremony was on the 11th. It began at the Gare du Nord, where imposing preparations had been made for the reception of the Unknown. In the presence of the King, the Crown Prince, State, military and court officials and heads of Allied and associated military missions, the procession started on foot to the Column of Leopold. Only one speech was made there and that by King Albert, after which General Allen pinned the Congressional Medal of Honor on the flag covering the casket, Other allied representatives also decorated the Unknown with medals.

The religious ceremonies took place at the Church of Sainte Gudule and were most impressive, They were attended by the royal family, Court, diplomatic corps, military missions, Senate and Chamber.

Prominent American Visitors.

General Clarence C. Williams, Chief of Ordnance, U.S.A., who was on an official tour of Europe to secure technical information relating to armament, visited the Commanding General in November.

Senator McCormick, of Illinois, was also a visitor.

Indication of Withdrawal of A.F.G.

A cable from Washington at the end of November announcing that no more replacements would be furnished for the A.F.G. indicated that the early departure of these forces was being seriously considered.

French Intentions in Ruhr.

M. Tirard informed General Allen on December 6th that France would probably go into the Ruhr District to secure additional guarantees for whatever was agreed on at the Brussels conference. General Allen informed M. Tirard that, in his personal opinion, such a move would not be sympathetic to American public thought.

Incident pertaining to Dual Command.

General Semaire, General Degoutte's new Chief of Staff,

called to pay his respects on December 11th and General Allen took occasion to give him the details of certain incorrect acts of the French commander of troops in the American zone. The Commanding General later had occasion to send for Colonel Piazza on account of several communications of an insubordinate and impolite character which the latter had addressed to Headquarters, A.F.G. Later, in conference with the French General in the American zone, it was decided that Colonel Piazza would be more useful in some other area.

The Commanding General had a long talk with M. Tirard about this time, during which he again made it clear to the Commissioner that the maximum of French show in the centre of the American zone had been reached. General Henrys, commanding the corps of which the French troops in the American area were a part, conferred with General Allen on the Piazza incident a few days later. General Allen did not insist on Colonel Piazza's removal, but it was agreed that he should write an apology to General Allen.

<u>Instructions on American Administration.</u>

Owing to a difference of viewpoint between some of the civil and military officials, it became necessary to emphasize the importance of having the American zone conform to the administration of the whole. This the Commanding General did in a communication issued to all officers and officials in the area, which follows :

<u>C O N F I D E N T I A L.</u>

HEADQUARTERS
AMERICAN FORCES IN GERMANY
COBLENZ, GERMANY.

4th January, 1923.

To : Chiefs of Sections and Services.

INSTRUCTIONS SUPPLEMENTARY TO THE LETTER OF 14TH JAN. 1920 DEFINING THE RELATIONS OF THE MILITARY AUTHORITIES TO THE INTERALLIED RHINELAND HIGH COMMISSIONS.

1. Inasmuch as the Treaty of Peace between the United States and Germany has come into force since issuing the original instructions defining the relations of the Military to the Interallied Rhineland High Commission, the following additional and supplementary instructions are issued for the guidance of all concerned.

2. Due to the fact that the United States Senate declined to ratify the Treaty of Versailles or the Rhineland Agreement dependent upon it, there is perhaps a thought that the Commanding General, American Forces in Germany has the powers in the American area of a general commanding an army of occupation during a state of war; that the "Rules of Land Warfare" lay down the principles applicable to the exercise of these powers; and that all officials who have functions to perform in the American area perform these functions subject to the war powers of the American commanding general.

3. Attention however is invited to the fact that a state of peace has existed between the United States and Germany since November 11th, 1921. The "Rules of Land Warfare" are mostly applicable to a state of war and not to a state of peace. The powers which may properly be exercised since the entry into effect of the Treaty of Peace, must be derived from that treaty.

4. Officials of other nations are charged by their respective governments with the exercise of certain functions in these territories. They also derive their powers from a treaty with Germany. These governments have not renounced, in the interest of the United States, any of the powers derived from their treaty. They are not parties to the treaty between the United States and Germany. The treaty between the United States and Germany became effective November 11th, 1921. The treaty between Belgium, France and Great Britain on the one side, and Germany on the other, known as the Treaty of Versailles, and the Rhineland Agreement, which was entered into by virtue of Article 432 of the former, both became effective January 10th, 1920. The Rhineland Agreement establishes the Interallied Rhineland High Commission, and defines its powers and functions. The pertinent provisions are as follows :

"Article 2.

"There shall be constituted a civilian body styled the Interallied Rhineland High Commission, and hereinafter called the High Commission which, except in so far as the Treaty may otherwise provide, shall be the supreme representative of the Allied and Associated Powers within the occupied territory."

"Article 3.

"(a) The High Commission shall have the power to issue ordinances so far as may be necessary for securing the maintenance, safety and requirements of the Allied and Associated forces. Such ordinances shall be published under the authority of the High Commission, and copies thereof shall be sent to each of the Allied and Associated Governments and also to the German government.

When so published they shall have the force of law and shall be recognized as such by all the Allied and Associated military authorities and by the German civil authorities.

"Article 5.

"The civil administration of the provinces * * * * shall remain in the hands of the German authorities, and the civil administration of these areas shall continue under German law and under the authority of the Central German Government, except in so far as it may be necessary for the High Commission by Ordinance under Article 3 to adapt that administration to the needs and circumstances of military occupation."

5. It is true that the United States is not a party to this treaty and is not bound by it and therefore loses no rights by its application. It is equally true that Belgium, France and Great Britain are not parties to the American treaty and are not bound by it and likewise lose no rights by its application, but the American government claims that none of the rights accruing to the nations who shared the burdens of the war are lost as a result of the Treaty of Versailles.

The treaty of the United States with Germany, properly considered, provides, insofar as Germany and its citizens are concerned, large governmental powers to the American commanding general. It is undoubtedly true that the treaty of Belgium, France and Great Britain with Germany, and the dependent Rhineland Agreement, properly construed, provide insofar as Germany and its citizens are concerned, large governmental powers to the Interallied Rhineland High Commission. No treaty or other instrument defines the powers of these authorities of occupation as between themselves.

6. On the request of the State Department the government of the United States authorized the Commanding General, American Forces in Germany to act as the unofficial American representative on the Interallied Rhineland High Commission. This has avoided conflict of authority. No ordinances, orders or decisions of the High Commission are promulgated until after

they have received the consideratiln and approval of the Commanding General and they then become effective in the American Area by his orders. The policy of the Commanding General is to cooperate with the Interallied Rhineland High Commission in every practicable way and a decision not to apply, in the American Zone, an ordinance of the High Commission, is to be regarded as a serious matter, not to be taken except for grave reasons,

7. From the foregoing, the policy of the Commanding General, and the reasons on which it is based, should be clear to all concerned, and there should be no doubt as to the action to be taken in any except extremely delicate or complicated instances. Any officer finding himself in this latter situation, should, in order to avoid an unintentional failure on his part to carry out the announced policy of the Commanding General, bring the matter, through his military superiors, to the attention of the Commanding General for final decision.

By Command of Major General ALLEN :

WM. W. HARTS,
Chief of Staff.

Reviews, Inspections and Sports.

The rifle and machine-gun teams representing the American Forces in Germany at the Interallied Shoot at Cologne on October 6th made a clean sweep, winning every event and the competition by a wide margin. The fact that the Allies had tens of thousands to pick from where the Americans had hundreds only, made the victory more impressive.

The Commanding General and the Chief of Staff inspected the 8th Infantry on October 11th. After the inspection and review, General Allen presented the winners in the various annual A.F.G. small arms' competition with trophies and medals.

The American team was successful again at a second Interallied small arms' competition, winning first place in both the officers' and enlisted men's class.

The Commanding General and the Chief of Staff were present at the review and inspection of the entire British

Army of the Rhine, held at Cologne in honor of Lord Derby, British Secretary of State for War, on December 14th. Besides General Allen, the Commanding Generals of the Belgian and French Rhine Armies were present.

Battlefield Trip.

A party of twenty-one A.F.G. officers made a battlefield trip early in December which took them to Verdun and other battlefields, including those of the Franco-Prussian war, in the vicinity of Metz. French officers familiar with the ground and the engagements fought there acted as guides for the party.

Holidays.

Thanksgiving Day, the fourth on the Rhine, was fittingly observed throughout the American area. No effort was spared by the messes to serve the troops the best Thanksgiving dinner that could be cooked.

The Commanding General sent holiday telegrams to, and received messages from, all the chiefs of the Allied forces on the Rhine on December 24th.

High Commission Forbids Posters.

The High Commission passed an ordinance on December 30th, forbidding the display in occupied territory of posters purporting to be extracts from a speech by Mr. Lloyd George derogatory to the French policy of "productive pledges," notably the occupation of the Ruhr.

JANUARY, 1923, UNTIL END OF AMERICAN REPRESENTATION.

Failure of Paris Conference.

The conversations on the reparations problem were resumed at Paris on January 2nd, 1923. The Allied governments of Belgium, France and Great Britain were represented by their respective Prime Ministers, with the exception of Italy, in whose behalf Signor della Toretta negotiated, Premier Mussolini having elected to remain at home. The German government dispatched State Secretary Bermann to Paris in order to represent the German case if necessary. However, although several proposals were studied closely, no agreement could be reached. Consequently, the conference broke up in "complete" but "friendly" disagreement on January 4th.

The principal point of difference was that the British government believed that the French decision to seize guarantees in Germany would seriously interfere with any attempts by Germany to reform its finances and with the possibility of obtaining large reparations payments in future. The British plan provided for a reduction of the German reparations debt to 50,000,000,000 gold marks and for a moratorium extending over several years, without resorting to the pledges contemplated in the French program. The British Prime Minister proposed further that Germany accepó financial control by a committee composed of representatives of Great Britain, France, Belgium, Italy, the United States and a European neutral nation. The Belgian and Italian ministers supported M. Poincaré's position, thereby authorizing the French government to make preparations for putting its plan of productive pledges into effect.

In commenting on the abrupt end of the conference, the German government pointed out in an official statement that no German proposals had been submitted because of the rejection of the British standpoint. The statement added that, in spite of this new disappointment, the German government would not give up the reparations program laid down in its note to the Reparations Commission of November 13th, 1922.

Germany's Voluntary Default Established.

After hearing German representatives on January 9th, who endeavored to prove that Germany's alleged delinquency in the coal deliveries was the consequence of natural causes for which it was not responsible, the Reparations Commission nevertheless established a voluntary default on the part of Germany, stating that it had not delivered the prescribed quantities in 1922. Sir John Bradbury, British delegate on the Commission, voted against the decision and Mr. Boyden, American unofficial representative, advised against it.

Apparently unable to agree in its session of January 13th, the Reparations Commission extended from January 15th to January 31st the period within which Germany was required to make certain payments, thereby postponing its decision regarding the German request for a moratorium. Two other voluntary defaults were established by the Commission on January 15th, immediately after the announcement by the German government that no more reparations deliveries to France or Belgium would be made and that cattle deliveries would likewise cease.

Finally, in its session of January 26th, the Commission established a general voluntary default by Germany with regard to its reparations payments under the clauses of the Versailles treaty. The British delegate abstained from voting. The Commission notified Germany of the decision made, at the same time stating that there was no need of a reply to the government's request for a moratorium and that, under these circumstances, the London plan of payments of May 5th, 1921, would become effective again. The note was signed by the French and Belgian members only.

Announcement of Sanctions in Ruhr District.

The Reparations Commission having formally established Germany's voluntary default, a Franco-Belgian note informed the German government on January 10th that sanctions would be taken in the Ruhr District. It was announced that a mission of control, consisting of engineers, would be sent thither, accompanied by the necessary troops as a guard and to assure the execution of the mission's task. This was stated to be, to watch over the production and deliveries of coal and assure stridt application of the program drawn up by the Reparations Commission. The note stated expressly that the French and Belgian governments contemplated neither military operations nor an occupation of a political nature.

Extent of Ruhr Occupation.

The occupation of the Ruhr District began on January 11th, when the city of Essen was entered by French troops. Roehum was entered two days later, while Dortmund was occupied on January 16th. After the occupation of the city of Luenen on January 17th, no further troop movements into unoccupied Germany took place. At the end of the month, the line of occupation ran from Dusseldorf via Metmann and Hagen to Unna, the two last named towns remaining free from occupation, and, turning north, reached the outskirts of Hamm and further west, the Dortund-Ems Canal. The northern border line of the newly occupied territory ran north of the cities of Recklinghausen and Buer to Dinslaken. Thus the entire Ruhr District, the industrial heart of Germany, came under the absolute control of France and Belgium.

Proclamations of German Government.

Immediately after the beginning of the Franco-Belgian action in the Ruhr, President Ebert and the national cabinet issued proclamations to the German people protesting against the measures. It was pointed out that Germany had not the military power to prevent the invasion by force of arms, but that everything would be done to shorten the period of occupation and alleviate the suffering of the population affected. The people were urged to refrain from irresponsible acts and to stand united with the entire German nation in the hour of distress. The proclamation further lodged a solemn protest to the whole world against this "violation of the peace treaty by France and Belgium". In special statements given to the press, both the Chancellor and the Foreign Minister called the Ruhr occupation a "breach of justice and act of violence against a defenceless nation", which was based on nothing in the peace treaty. This view was shared by all the political parties, ranging from the G.N.P.P. to the United Socialists.

A German Note of Protest.

The national government delivered a note to the French and Belgian Ambassadors at Berlin on January 12th in reply to the Franco-Belgian communication of January 10th. It declared that it would "tear the mask from the face of the French government" and reveal the character of the proceedings in the Ruhr District in their true light. The decisions of the Reparations Commission did not jutify the Franco-Belgian action, which rather constituted a flat violation of the Versailles treaty and of international law. Refuting the French and Belgian statement on the peaceful character of the action, the note pointed out that armed contingents of troops had crossed the border of the occupied territory in a warlike manner and invaded Germany, and that the entire Franco-Belgian procedure was a purely military act. Finally, it announced that Germany would not be in a position to make payments or deliveries to France and Belgium while the illegal status in the Ruhr continued nor until the serious consequences thereof had been overcome.

Reichstag's Support of Government.

The Reichstag met in plenary session on January 13th for the purpose of defining its attitude towards the government's action in the crisis. The House showed surprising unity when the question was put to a vote. The measures taken by the government in protest against the occupation were approved by an overwhelming majority, only twelve deputies dissenting. Chancellor Cuno, addressing the Reichstag, reviewed the events which had led up to the occupation and repeated the statement that it constituted a violation of the Versailles treaty, since it lacked the joint approval of all the Allied governments. He announced that all payments and deliveries to France and Belgium would remain suspended until the "illegal status" in the Ruhr District had come to an end. Dr. Stresemann, speaking for the Bourgeois parties, promised to support the government. Deputy Hermann Mueller, on behalf of the United Socialists, approved the government's measures against the Franco-Belgian action, and Prussian Premier Braun declared in the name of all the federal states that their governments were unanimously in accord with the national cabinet's policy and that they would support it.

Protest Demonstrations.

Protest demonstrations against the military occupation of the Ruhr District were staged throughout Germany, the occupied area included, on January 14th and 15th. On the day last named, work was suspended for half an hour in all public works, factories, offices, banks, shops, restaurants, etc., people in the street stopping for two minutes "to reflect on the shame and violence that had been perpetrated." The restlessness of the population, encouraged by the government, became great. The situation created by the occupation had the effect of establishing a united front of practically all political parties from Right to Left, at least for the time being.

German Counter-Measures.

Upon receipt of the Franco-Belgian note announcing the occupation of the Ruhr District and the control of coal production, the national government recalled its ambassador at Paris and its chargé d'affaires at Brussels without, however, having the intention of breaking off diplomatic relations thereby. Furthermore, the National Coal Commissioner issued instructions to all the mines in the Ruhr District to deliver no reparations coal to France or Belgium, since the national government would no longer pay for any coal delivered to those countries or confiscated by them for reparations purposes. At the same time, the German War Burdens Commission at Paris informed the Reparations Commission that all reparations payments to Belgium and France were to be suspended. The German Coal Syndicate, the head office of which was at Essen, transferred its activities to Hamburg on January 10th. This step proved to be a great inconvenience to the French and Belgians with regard to the solution of the important questions pertaining to the general organization of the Ruhr District and the distribution of coal.

Passive Resistance.

The situation in the Ruhr District became very grave on January 15th, when the National Coal Commissioner expressly forbade any person or persons to deliver coal or coke to France or Belgium even in case of payment by those countries. In compliance with these orders, the mines suspended deliveries at once. Although the French immediately directed the mine owners to resume deliveries, threatening the application of serious sanctions at the same time, both the industrials and the labor unions maintained their obstinate attitude and declared that they would continue to obey the government's orders.

Reaction Upon Population.

Franco-Belgian interference in the administration of mines, railways and telegraph and post offices in the Ruhr

District resulted in partial strikes of miners, railwaymen and other employees, which lasted until the French measures were revoked. Railway traffic, however, continued to be abnormal because of interferences therein on the part of the military authorities of occupation. The sudden arrest of prominent mine owners and managers of mining concerns, including Herr Thyssen, on January 20th, incrdased the tension. A general strike was averted only by the release of Herr Thyssen and the other industrials, who were let off with heavy fines imposed upon them by a French court-martial at Mayence on January 24th. Their return trip from Mayence to Essen resembled a real triumphal progress, huge demonstrations being staged in their favor at all the stations passed by the train. It was reported that they had decided to request the Paris Court of Appeal to revise the sentence.

Results of Occupation.

When the month of January approached its end, hardly any coal or coke had been transported to France and Belgium since the first days of the occupation as a result of acts of sabotage and passive resistance on the part of German employers and employees. On the other hand, the supply of the unoccupied territory was increased considerably during the last days of January. In order to stop further consignments of coal into unoccupied Germany, the French and Belgian authorities encircled the entire Ruhr District by a cordon of troops and a customs line on February 1st, thereby establishing strict control of all outgoing trains.

Germany's Firm Attitude.

There was a lively exchange of notes between the French and German Governments in connection with the measures in the Ruhr District, especially the arrests of the coal magnates and certain government employees, and more demonstrations against the occupation were staged in the Prussian State Assembly and the Reichstag on January 25th, both Houses again protesting against the Franco-Belgian act. Speaking on the national budget and the political situation, Minister of Finance Hermes repeated his former statements that neither the German government nor the German people would surrender to force and that no reparations negotiations could take place under the Ruhr District be evacuated. The Minister emphasized the fact that the nation's existence was at stake.

Relief for Ruhr Population.

A nation-wide movement for assistance of the Ruhr population set in shortly after the occupation of the Basin. Collections which yielded substantial sums were made by national, State, municipal and private organizations. Funds were also placed at the disposal of the National President by foreigners and Germans living abroad who desired to express their sympathy with the people of the occupied district.

Monarchist Demonstrations in Bavaria.

The foregoing situation occupying the centre of the political stage, very little heed was paid to domestic questions during the month. The National Socialists (militant monarchists) of Bavaria once more came into the limelight by compelling the Bavarian government to proclaim a state of martial law in Bavaria on January 26th as a result of the convention of this reactionary party which took place at Munich on January 27th. Apparently yielding to the pressure of the National Socialists and the Bavarian Bourgeois parties in general, the Bavarian government finally modified its decision and permitted the convention to be held. With the exception of encounters of minor importance between Communists and members of the Hitler party, no incidents were reported in connection therewith.

Separatist Movement in Hanover.

In view of the occupation of the Ruhr District, the Hanoverian Separatists (Guelphs) announced that they had given up their intention of holding a referendum on the question of separation from Prussia at the present juncture.

Invasian of Memel Territory.

While the Ambassadors' Conference was still deliberating on the fate of Memel, armed Lithuanian irregulars, organised on the pattern of the Italian Fascisti, succeeded in occupying that territory, including the city of Memel, on or about January 12th.

Exchange Rate.

Germany's entire economic situation in January, 1923, was dominated by the abnormal conditions prevailing in the Ruhr District, the economic heart of the nation, as a result of the Franco-Belgian occupation. The acute economic and political crisis caused by the abrupt end of the Paris conference on January 4th, and the fateful consequences thereof, made themselves felt by a further fall of the German exchange rate, which had remained comparatively steady during the month of

December, 1922. This unprecedented decline, which continued throughout the month of January and early in February, is illustrated by the following quotations of the Cologne Bourse :

	Jan. 5th.	Jan. 10th.	Jan.27th.	Feb.3rd.
American dollar	8,578.60	101648.31	27,461.17	37,506.00
British pound	39,376.31	49,002.18	126,458.06	174,462.25
French franc	590.52	716.18	1,770.56	2,443.87
Dutch guilder	3,371.55	4,244.36	10,852.80	14,942.25
Belgian franc	545.13	657.81	1,551.11	2,094.75

National Finances.

The precarious condition of the national finances at this time was illustrated by the official government and Reichsbank statements. Germany's floating debt reached the huge total of 1,821,616,367,000 marks on January 20th. Minister of Finance Hermes declared in the Reichstag on January 25th that it had increased to 1,852,000,000,000 marks by January 22nd. In this connection, he pointed out that it had risen from 51,000,000,000 to 166,000,000,000 during the period ranging from the end of the war to December 31st, 1920, alone, and that the increase in the year 1922 and during the first two ten-day periods of January, 1923, amounted to 1,500,000,000,000 marks.

In the same speech, the Minister stated that the deficit of the national budget was about 1,600,000,000,000 marks - figures which approached those common in Russia. It was impossible to draft a proper budget plan in this disorganized state of finances, he said, and he blamed the Franco-Belgian action in the Ruhr District for these disastrous conditions.

Cost of Living.

In consequence of the fall of the mark, wages, prices and the cost of living increased from day to day throughout the month. The additional charge on customs duties was fixed at 3,230 paper marks = 1 gold mark at the end of the month. The index figure for wholesale trade, as calculated by the National Office of Statistics, was 2,131 on January 15th and increased to 3,286 or 54%, on January 25th. At the same time, the index figure for food prices rose from 1,703 to 2,764, or 62%, while that for imports increased from 3,316 to 5,360, or 62%. The cost of living for a workingman's family of two adults and two children for a period of four weeks was estimated by the Cologne Office of Statistics at 190,018 marks on January 25th, in comparison with 120,024 marks on January 10th.

The increase in the cost of living made itself felt especially in the Ruhr District, where prices of the most important foodstuffs increased by 46.3% from January 11th to 20th.

Fuel Situation.

Immediately after the occupation of the Ruhr District, a bitter struggle ensued between the occupying authorities and the German mine owners, industrials and operators. The latter adhered throughout the month to instructions issued by the German government, which ordered the discontinuance of all consignments of reparations coal to either Belgium or France. Partial strikes, caused by temporary military occupation of certain mines, and the general abnormal conditions in the Ruhr District, certainly reduced the coal production considerably, although no statistical figures were available at the end of the month. The French and Belgian authorities succeeded in diverting no coal trains to France or Belgium as a result of the recalcitrant attitude of the Germans, ranging from the mine owners to the last miner, all of whom abandoned their former political and economic grievances and formed a "united front of defence" in unanimous support of the national government. On the other hand, the occupying authorities were unable to prevent local consignments to unoccupied Germany up to the end of January, but succeeded in changing these conditions on February 1st by the establishment of the customs line on the eastern border of the district to which reference has already been made.

Contracts with Foreign Firms.

In anticipation of a suspension of further coal consignments to unoccupied Germany, the German government and industrial magnates consequently concluded contracts for British coal with British firms. According to the "Weekly Dispatch", orders for 200,000 tons were placed by the Germans, delivery to be made by the latter part of May. In view of the continued decline of the mark and increase in wages and salaries, the National Coal Council decided on a considerable rise of coal prices, which became effective on February 1st.

Transportation.

Germany's transportation facilities in January were hampered by the various railway strikes in the Ruhr and other parts of occupied territory, as well as by the increasing shortage of coal. It being reported in January that the coal supply would last only about one month more, the National Railway Administration decided to restrict passenger traffic by 10% or 20%.

Economic Warfare.

With regard to the Franco-Belgian occupation of the Ruhr District, German economists expressed the view that the action was bound to be unsuccessful because of the engergetic resistance on the part of the German government, industrials and laboring classes. They further predicted that the final outcome of the coal struggle, continued by the establishment of the customs border and the separation of the Ruhr from the remainder of Germany, would likewise be a negative one. Considerable importance was attached to the coal agreements between German and British financial circles and to another counter measure, viz., the conclusion of contracts with Swedish concerns for delivery of Swedish ore during a period of ten years. All the current contracts between the French ore mining industry in Lorraine and German smelting concerns were cancelled. When the American Army left the Rhine, an economic "war to the knife" had already broken out between France and Germany, destined, perhaps, to decide the future of both countries.

In Occupied Zone

Reaction of Ruhr Occupation in Rhineland.

The occupation of the Ruhr District naturally caused much alarm and deep indignation among the population of the occupied Rhenish territory. Although with certain restrictions ordered by the occupying authorities, the provisions of the program of national mourning were carried out by the population on January 14th and 15th. Both days passed off without incident.

I.A.R.H.C. Opposes German Government.

The German Government having given orders to suspend deliveries in kind to France and Belgium in consequence of the entry of French and Belgian troops into the Ruhr. M. Tirard, President of the High Commission, thereupon forwarded instructions to the Kreis delegates in the French and Belgian zones of the Rhineland, stating that the German order would not apply to the occupied territory, because it had not been submitted to the High Commission for approval ; further, that officials who complied with it would incur administrative and judicial penalties as laid down in the ordinances.

The recalcitrant attitude of the German Government caused the Interallied Rhineland High Commission to issue three ordinances on January 18th providing for the adoption of new sanctions and pledges, namely, introduction of coal tax, confiscation of customs receipts and exploitation of fiscal forests throughout the occupied territory. The German government instructed its diplomatic representatives at Paris, Brussels, London and Rome on January 22nd to deliver a note of protest against these ordinances to the governments to which they were accredited. The note stated that the I.A.R.H.C. had committed a grave violation of the administrative and financial sovereignty of the German nation and the federal states concerned by issuing these ordinances, and had assisted the Franco-Belgian military action in the Ruhr, which in itself was an "infringement of the treaty of Versailles and of international law". It further declared that the instructions embodied in these ordinances to German officials and government employees to cooperate in the execution of the measures were an unheard-of demand and that the government had informed its officials and employees in the Rhineland that all orders which constituted an infringement of the Rhineland agreement were ineffective and must not be complied with by them.

Arrests and Expulsions of German Officials.

The consequence of the attitude of the German officials and employees in the Rhineland who elected to support their government and refused to obey the orders of the occupying authorities was the arrest or expulsion from the occupied territory of a large number of these officials, especially in the French and Belgian zones. The total number of officials expelled from the original occupied territory was estimated to have reached 150 early in February.

Martial Law.

The result of these expulsions of many high ranking government officials was a strike in several cities of the occupied territory, accompanied by street demonstrations and riots. For this reason, a state of martial law was proclaimed by the French authorities at Treves and Kraeznach on January 25th, while the same measures were taken by the Belgians in certain districts of the Lower Rhine. The military occupation of certain railway stations by the French resulted in a standstill of practically the entire railway service on the left bank of the Rhine on January 27th.

German Protests Against Expulsions.

Several notes were delivered by the German government to the governments of Great Britian, France and Belgium, protesting against the expulsion of German officials from the occupied territory and the system "aimed at compelling German officials to act against their country." The French government replied on January 24th and 25th that those measures were only a consequence of Germany's violation of the treaty and that they had no purpose other than to guarantee the execution of that pact. The French government stated further that the German government's decree forbidding its officials to obey the ordinances was in itself a violation of the treaty. M. Poincare added that this note constituted a reply in advance to any later messages which might be sent by Germany to the French government on the same subject.

Separatism.

The Rhenish Separatists made the most of the occupation of the Ruhr District by increasing their propaganda activities throughout the old occupied territory as well as the Ruhr District itself. On the other hand, anti-Separatist propaganda was also carried out on a large scale. Pamphlets containing proclamations to the Rhinelanders by the united political parties of the Rhineland and directed against the machinations of Herr Smeets, leader of the Rhenish Republican People's Party, were issued with copies of anti-Separatist newspapers. The statements and promises made by the Smeets Party, especially with regard to the introduction of the so-called "Rhenish franc", were energetically opposed and refuted. Anti-Separatist demonstrations occurred in several cities of the occupied territory at this period.

In American Area.

Ruhr Occupation and Withdrawal of American Forces.

The Franco-Belgian occupation of the Rhenish-Westphalian industrial district and the withdrawal of the American Forces from the Rhineland, which was ordered on January 11th, engaged the political attention of the population of the local area practically throughout the month. While the unreserved condemnation of the Franco-Belgian step by the population and the newspapers of all political shades was unanimous, the effect of the announcement of the withdrawal of the American troops was one of regret. The view was expressed generally that as long as any occupation existed, the presence of a small force of American troops would be very much in Germany's interests for political reasons, inasmuch as it was held that the United States was disinclined to permit a change in the political status of the Rhineland. Fears were expressed by the entire German press that the peaceful settlement of the great reparations conflict and the final establishment of universal peace would be delayed as a result of the decision by the American Administration. The departure of the Americans from Coblenz on January 24th prompted the "Coblenzer Zeitung" (G.P.P.) to review the history of the American Forces in Germany during their four years' stay on the Rhine in a valedictory article. The paper stated that relations between them and the population had gradually become friendly, thanks to the policy pursued by the Commanding General and other responsible officers, who had made every effort to reduce the heavy burdens of occupation to a minimum. It also declared that the relations between Headquarters and the Coblenz press had been of a cordial nature, Headquarters having always been willing to lend assistance whenever necessary. The writer regretted the withdrawal of the troops at such a serious juncture, stating that the withdrawal was not in the interests of Germany, although it was true that it was doubtless an act of protest against the occupation of the Ruhr.

Further Expressions of Regret.

Prince Hatzfeldt, German National Commissioner for the occupied Rhenish territories, called on General Allen on January 18th and presented a letter from Chancellor Cuno, in which the German Chancellor expressed regret at the departure of the American troops and thanked the General in the name of the German government for the spirit in which affairs in the American area had been administered.

In his reply to Chancellor Cuno's letter, forwarded to Prince Hatzfeldt for transmission, on January 23rd, General Allen expressed the sincere wish that conditions in Central Europe might soon assume a more hopeful aspect.

Many French and German papers also expressed regrets at the withdrawal of the American troops and the termination of American influence on the Rhine.

General Allen's Mediatory Efforts in Ruhr Problem Vain.

On January 11th, the first day of the Ruhr occupation,

General Allen suggested a meeting of all the High Commissioners, hoping that they might collectively feel that changed conditions warranted an effort to renew negotiations with a view to finding a satisfactory arrangement between Germany and the Allies on the Ruhr question. This meeting, in which M. Tirard, Baron Rolin-Jacquemyns, Lord Kilmarnock and General Allen participated and which was called by the latter in a private capacity and not as an official designated by his government to undertake it, was held in M. Tirard's office on January 12th. Although all the Commissioners displayed a sympathetic attitude toward General Allen's action and agreed to bring the matter before their respective governments, yet they realized that a discussion of a renewal of negotiations between the Allies and Germany was not the business of the Commissioners, either individually or as a body.

General Allen informed the State Department on January 13th of his personal and unofficial representations regarding a renewal of negotiations on the Ruhr problem. Although he realised that the question was not properly within the competence of the High Commission, he believed that the seriousness of the situation and the extent to which the High Commission was concerned in the French plans would justify an expression of opinion by that body. He also realised that his was a bold step to take without previous instructions from the State Department, but felt that should any possible good result from it, the initiative would be justified.

General Allen received word from the State Department

on January 18th, indicating that his efforts to bring the extreme necessity of a renewal of negotiations before the several Commissioners had met with an unsympathetic reception at Washington. The General was instructed to take no action other than his essential military duties in connection with winding up the affairs of the A.F.G. It may be stated also that Lord Kilmarnock was instructed by his government to abstain from discussing matters at Commission meetings bearing on the French plan of action in the Ruhr District.

Lowering of American Flag on Fort Ehrenbreitstein.

The ceremony incident to the lowering of the American flag and the raising of the French tri-color at Fort Ehrenbreitstein on January 24th, the day of the departure of the American troops from Coblenz, was very simple and impressive. The American flag was actually lowered by one non-commissioned officer and two privates of the old 7th Machine-Gun Battalion and was caught by General Allen as it neared the base of the staff, a French band playing the "Star-Spangled Banner" the while. The flag was laid down by the General's aides and three enlisted men, and immediately thereafter the French flag was taken to the base of the flagstaff by three French soldiers. General Mordacq, representing General Degoutte, held it while it was being attached to the lanyards. As it was released to the breeze and was slowly raised over the old fort, the 8th Infantry band played the French national anthem. Following this, Companies D and M, 8th Infantry passed in review before General Allen and his staff, General Mordacq, the High Commissioners, Colonel Bates and the liaison officers attached to American Headquarters.

Anti-Separatist Demonstrations.

Separatism in Coblenz suffered a serious set-back on January 25th, when the printing officers of the Dorten press and the Rhenish Republican People's Association of Bad Ems, situated in the Schloss-strasse, were invaded by an angry crowd and the plant was entirely demolished. The organizers of the riot having succeeded in obtaining membership lists of the Dorten Press, the lodgings and business places of several alleged Separatists of Coblenz were ransacked on the same day. Due to the huge ovation given Herr Thyssen and the other mine owners at the Coblenz railway station, patriotic feeling ran very high in the city on that day. Another demonstration in front of the Separatist newspaper offices was staged after the reception of the coal magnates, but the crowd was dispersed by the German police. Due to proclamations by the local police administration and the political parties, warning the population against excesses, the excitement gradually died down and no other disturbances occurred during the month of January.

Protests against Expulsions.

Although no deportations of German officials took place in the local area in January, the united political parties and economic organizations of the electoral district of Coblenz-Treves protested jointly on January 22nd against the arrest and expulsion of German officials from other districts of the occupied territory. The appeal stated that the entire population supported these loyal officials and employees and that the people made no distinction whatsoever between Rhenish and non-Rhenish officials. In addition, the chiefs of all the national, state and municipal authorities of Coblenz adopted a resolution on January 22nd and submitted it to the I.A.R.H.C. They stated therein that they were unanimously determined to support their government and that they were bound by their oath and conscience to comply with its orders. They expressed the conviction that German and Prussian authorities should not be called upon to carry out orders directed against the interests of the German nation and incompatible with the terms of the Rhineland Agreement.

Change of Command in Local Area.

At midday on January 27th, General Allen turned over the command of the American area to General Marty, the French officer designated by General Degoutte to assume it. The ceremony was a simple one, only the officers of General Allen's staff and Captain Hendrickx, French liaison officer with the A.F.G., in addition to General Treat, a class-mate of General Allen, being present. General Allen thanked General Marty in a brief address for his cooperation as commander of the French troops in the American area and complimented Captain Hendrickx for the work done by him in the last four years. General Marty replied in kind. The entire ceremony lasted only a few minutes. The following proclamation by the Chief Burgomaster, which was posted throughout the city, informed the Coblenz population of this historic event:

"In the interest of the maintenance of peace and order, I wish to announce that the military command was transferred from the American authorities of occupation to the French authorities at noon today.

Coblenz, 27th January, 1923.

The Chief Burgomaster."

Proclamation of Railway Strike.

Immediately after this transfer of power, the German railway employees at the Coblenz main station and the Coblenz-Luetzel station went on strike because of occupation of those two stations by French troops. In the course of the day, practically the entire railway traffic on the left bank of the Rhine came to a standstill, while the traffic on the right bank was greatly affected as a result of the closing of the line running from Ehrenbreitstein in the direction of Niederlahnstein, service between Cologne and Ehrenbreitstein, however, being maintained. These conditions still existed on February 3rd, the German railway employees having persisted in their demand for withdrawal of the French military guards from the stations, signal-cabins, etc. The French attempts to conduct trains with French personnel were not very successful, due especially to lack of locomotives and to acts of sabotage.

French Precautionary Measures.

The local French authorities of occupation issued an order through the local police administration on January 30th prohibiting public gatherings of more than ten persons, as well as noisy demonstrations in the streets of the city.

Suspension of Newspapers.

Orders forbidding the circulation of certain newspapers were issued by the local delegate of the I.A.R.H.C. on February 1st. The two big Cologne dailies, the "Koelnische Zeitung" (D.P.P.) and the "Koelnische Volkszeitung" (Centrum), were prohibited in the city of Coblenz for a period of nine days beginning on February 1st, because they published protests by the local German authorities and officials against the occupation of the Ruhr and the Commission's ordinances in connection therewith. This publication, according to the delegate, was calculated to increase the resistance of the population to the authorities of occupation. In addition, the French District Delegate sent a notice to the Coblenz newspapers on February 1st with the request that they publish it immediately. The notice was an attempt to prove that German officials were not entitled to contest either the legality or illegality of the I.A.R.H.C.'s ordinances. The papers were requested to reprint the statement in accordance with Article 97 of the Rhineland agreement. After a careful study of the contents of the notice, which was received by all the Coblenz newspapers, the "Rheinische Rundschau" (Democrat) excepted, the papers arrived at the conclusion that they were not based on the Rhineland agreement, but were intended "to support the policy of pledges and hamper the orders of the German government." For this reason, the editors and the entire staffs of the Coblenz newspapers, including the "Rheinische Rundschau", declared themselves unable to comply with the I.A.R.H.C.'s demand. Consequently, all the local newspapers, with the exception of the "Rheinische Rundschau" were suspended on February 2nd for a period of three days. While the Separatist organs of the Dorten and Kaffine groups were unable to appear as a result of the demolition of their offices, the "Rheinische Rundschau", for the sake of solidarity, ceased publication voluntarily and did not appear as long as the circulation of the other Coblenz papers was suspended. The Chief Burgomaster, the political parties and trade unions lodged a protest with the local French delegate of the I.A.R.H.C. against the suspension of the papers.

Explusion of Ober-Praesident.

Dr. Fuchs, Ober-Praesident of the Rhine Province, was expelled from the occupied territory on February 2nd after a brief conference with M. Tirard, President of the I.A.R.H.C. This measure on the part of the I.A.R.H.C. resulted on February 3rd in a one-hour protest strike at Coblenz and a demonstration in front of the building occupied by the I.A.R.H.C. No particular incidents occurred, however, and the crowd was dispersed by French troops.

General Allen received a communication from Ober-Praesident Fuchs before his departure, expressing regret that he had not had an opportunity to call on the General and expressing his appreciation of the courtesy and consideration shown in the administration of the American area, saying that the departure of the American was a very great misfortune to Germany.

General Allen's Formal Withdrawal from I.A.R.H.C.

While conditions at Coblenz were thus assuming a critical form, General Allen, in accordance with orders from the State Department, formally withdrew as American observer from the I.A.R.H.C. on February 3rd. The event was marked by the lowering of the American flag from the Commission building on the same day.

New Year's Telegrams.

The Commanding General sent New Year's telegrams to the various American embassies in Europe and received greetings from many officials both in Europe and America.

Talk with Ober-Praesident.

The General paid a visit to Dr. Fuchs, Ober-Praesident of the Rhine Province, on January 2nd. During the hour in which they talked, Dr. Fuchs spoke very frankly about conditions in Germany and appeared particularly concerned as to the Separatist movement.

Imminent Occupation of Ruhr.

In a talk with M. Tirard on January 6th, the General learned that there was a bare possibility that an agreement would be brought about between France and Germany, but that in the event no agreement was reached, troops would be sent into the Ruhr.

Colonel Francois, a member of General Degoutte's

staff, called on General Allen on the 7th and delivered a letter from General Degoutte informing General Allen of a concentration of French troops preliminary to entering the Ruhr. General Allen informed the State Department by cablegram on the same day of the conditions in the Rhineland and the proposed measures to be taken by France. Information from reliable sources indicated that the French concentration would be quickly followed by an advance into the Ruhr, and that all political parties in Germany seemed solidly organized against the French move. Seventeen military trains passed through Coblenz en route to the concentration area on January 8th.

A.F.G. Recalled.

A cablegram from the War Department received on January 11th confirmed the report that the A.F.G. was to be withdrawn, and stated that the U.S.A.T. "St. Mihiel" making only one trip, would leave New York about January 12th for Bremen or Hamburg. The cablegram also outlined the policy to be followed in disposing of surplus supplies and routing of personnel in excess of the capacity of the transport. The recall of the A.F.G. coinciding with the advance of the French into the Ruhr, gave the impression that the American action was intended as a rebuff to the French policy. This impression was prevalent in all circles in spite of the announcement made at Washington that the action was not intended as a reflection on American friendship for France.

Letters were received from Field Marshall Robertson and Ambassadore Herrick expressing regret at the departure.

On January 22nd, General Allen informed General Degoutte of the details governing the departure of the American Forces and stated that he would be ready to turn over the command of the American area and the Coblenz bridgehead on January 27th or any date thereafter, to whomever General Degoutte might designate.

Invitations were issued on the 23rd to the High Commissioners to be present at the ceremony of the lowering of the American Flag at Fort Ehrenbreitstein, which had been fixed for the 24th. This brought a joint response from the Belgian and British Commissioners, saying that the departure of the American troops and the lowering of the flag was a source of such keen regret to them that they begged to be excused from witnessing the ceremony. They added that they would be present at the station when the troops left.

General Allen's Final Message to A.F.G.

General Allen issued his final message to the A.F.G. on the 24th. This appeared in the last issue of the "Amaroc News" and was as follows :

> Today at noon the American flag will be lowered on Ehrenbreitstein, thus making the end of the occupation of the Rhine province by the American troops.
>
> For four years the forces of the American Army have held their place on the Rhine beside their Allies and have earnestly striven towards restoring peace under the terms of the adopted treaties. With deep affection in our hearts for our Allies and sympathy for our former foes, our highest ambition has been to act with such justice towards all as would insure a lasting peace in Europe.
>
> Since the departure of the Third Army on July 2nd, 1919, the American Forces in Germany have occupied the American zone, including the Coblenz Bridgehead, which they have con-

tinuously supervised and controlled. All through this period there has been a consistent effort on our part to execute this important task with a maximum of effectivemenss and a minimum of hardship on the population. My instructions to the command with regard to their correct personal conduct and considerate attitude towards the local population have been rewarded. The correctness of your conduct in general and the high standards of military efficiency set by the various organizations have made it possible to leave behind us an enviable record for justice, fairness and military excellence.

I am confident that this spirit of justness and fairness applauded by our Allies will in the days to come prevent any feelings of bitterness or rancor towards us in the minds of those among whom we have lived, and will mark a forward step in the accomplishment of a genuine peace. It is therefore with much pleasure and pride that I congratulate the officers, soldiers and the attached civilian personnel of this command on the part you have taken in the stirring events of the last four years, and particularly on the eminently successful execution of your mission. You have deserved well at the hands of your country and may always carry in your hearts the consciousness of a duty well done.

HENRY T. ALLEN,
Commanding General.

Departure of Troops.

The passage of the departing troops to the station between lines of French infantry on the 24th was very affecting. They were drawn up and inspected by the Commanding General, assisted by the High Commissioners and French generals, for the last time, after which they were marched on to their two train sections. All left on schedule time. The outstanding feature of the event was the genuine sorrow depicted on the faces of both Germans and Allies.

Head of German Military Commission Makes Final Call on General Allen.

Major Kuehlenthal, head of the German military mission, who had been associated with American Headquarters practically throughout the entire period of occupation, made his final call on the Commanding General on the 27th.

---- o o O o o ----

CHAPTER II

ACTIVITIES OF INTERALLIED RHINELAND HIGH COMMISSION.

Year 1922 — Page

Year 1923

----- ooo O ooo -----

ACTIVITIES OF INTERALLIED RHINELAND HIGH COMMISSION

IN YEAR 1922.

Separatist Movement.

A very far-reaching decision taken by the High Commission in 1922 was that whereby the German authorities were forbidden to execute certain sentences pronounced against Herr Joseph Smeets, leader of the Separatist movement in the Rhineland, for slanderous attacks on German officials.

The Separatist movement began during the period of the armistice, with Dr. Dorten as its leader. When, in 1919, an attempt to establish a Rhineland republic, with the connivance of the French authorities of Occupation, met with failure, Dr. Dorten ceased his activities for some time.[1] The Separatist movement was quiescent until 1921, when a certain Joseph Smeets began to revive it. He established a newspaper called the "Rheinische Republik", presumably with capital furnished by the French authorities. The purpose of this journal was to wean the population of the Rhineland away from Prussia and to encourage the separation of the Rhineland from Prussia and even from the rest of Germany, Smeets made the mistake of writing slanderous articles concerning certain German officials, including the President of the German Republic. The German authorities took action against him because of these articles, which were extremely insulting.

---

[1] A.M.G. of O.G. 1918, Vol. I., chapter 15.

As soon as Smeets was brought to trial he appealed to the High Commission, invoking protection under Ordinances 27 and 70, which grant immunity for political acts committed during the period of the armistice. Although the acts of which Smeets was accused were committed in 1921 and 1922, he claimed the protection of the High Commission, alleging that he was being prosecuted because of acts connected with a policy inaugurated during the period of the armistice and continued after the coming into force of the peace treaty.

When Smeets first appealed to the High Commission he found support with the French High Commissioner only, who asserted that French public opinion would not permit Smeets to be punished. The French High Commissioner claimed that Smeets was entitled to protection not only in virtue of Ordinances 27 and 70 but also in virtue of Ordinance 90, which was enacted to protect persons prosecuted for services rendered the authorities of occupation or because of relations with those authorities. At a later date, the Belgian High Commissioner, acting under instructions from his government, supported the French High Commissioner in his position and proposed that the German authorities be informed that they must drop all proceedings against Smeets. General Allen and the British High Commissioner did not concur in this view because they were of opinion that while the motive for Smeets' acts might be connected with what took place during the armistice, the acts themselves were separate acts occurring long after the armistice and that he therefore could not be protected under Ordinances 27 and 70. They were also of opinion that Smeets' relations with the authorities of Occupation did not warrant intervention of the High Commission. They expressed the view that he had rendered no services to these authorities and accordingly was not entitled to their protection. In view of the importance of this matter with respect to public order and administration in the American zone, General Allen set forth his views to the High Commission in a memorandum.[1] As a number of cases against Smeets were still pending, the High Commission, after much discussion and argument, decided not to object to the continuation of proceedings but to require the German authorities to await its decision before putting any sentence into effect.

In August, 1922, the German Commissioner for Occupied Territories informed the High Commission that the proceedings against Smeets had come to an end and requested that the sentence imposed by the German courts, which amounted to eight months and two weeks of imprisonment, might be put into execution. When this request came before the High Commission, the British High Commissioner stated that his government considered the Smeets case a most important issue, because a decision in favor of Smeets would constitute an unwarranted interference in German administration and might be regarded as an intention to intimidate German courts in the proper performance of their duties. He stated that if the High Commission extended its protection to Smeets

---

[1] See page 206

in the cases under discussion, such a decision might be considered by Smeets and his followers as a license to violate German law with impunity. He was of opinion that such a decision migh have the gravest consequences, in that it migh lead to the belief that the German authorities in the occupied territories had no right to punish anybody connected with the Separatist movement, irrespective of the nature of the offense committed.

The French High Commissioner argued that a decision in Smeets' favor could be justified on legal grounds, because it was clear that Smeets was simply continuing a campaign which had been begun during the armistice. This remark would lead to the belief that the arguments advanced by Smeets had been furnished by the French Legal Department of the High Commission. The French High Commissioner placed considerable stress upon the political side of the question because his government had unofficially declared that it would not permit Smeets and the other leaders of the Separatist movement to be punished. The Belgian High Commissioner stated that he thought the German authorities themselves responsible for the unfortunate situation which had been created, because they persisted in bringing petty charges against Smeets, whereas they allowed worse offenders to escape with impunity. He contended that none of the offenses of which Smeets was convicted was serious. In this respect, it may be remarked that Smeets had openly insulted the President of the German Republic and publicly advocated treason. The Belgian view was that the German authorities had deliberately put the High Commission in a position in which it must take action in favor of Smeets or suffer a very considerable loss of prestige, because if Smeets were not protected the Germans would point to the fact that a person who had had relations with the authorities of occupation had been punished in spite of those authorities. The High Commission therefore could have no other recourse than to order that Smeets should not be punished.

The French High Commissioner realised that it would not be possible to persuade General Allen and the British High Commissioner to agree to protect Smeets. He therefore proposed that the High Commission should simply inform the German authorities that the judgments given against Smeets should not be executed until further orders. General Allen and the British High Commissioner did not agree to this proposal, because they believed that Smeets and his followers would construe such a decision as a permission to defy the German authorities, and that this would create a condition for which the High Commission would be responsible. General Allen expressed the view that the Separatists might assume that the immunity granted Smeets extended to them and that it was therefore possible that serious disturbances of public order, which would involve the authorities of occupation, might take place in the occupied territories. In reply to this, the French and Belgian High Commissioners asserted that the trifling nature of the offenses of which Smeets was accused did not justify the belief that he or his followers would proceed to violence.

In spite of the arguments of General Allen and the British High Commissioner, the High Commission decided by a majority vote to sustain the appeals lodged by Smeets. This action was based on the assumption that Smeets had been prosecuted by the German authorities for a large number of acts, the majority of which were of a minor character and would not have led to prosecution unless account had been taken of the personality of their author and the political campaign conducted by him against the Prussian government during the period of the armistice; that Smeets had publicly manifested the desire to maintain good relations with the authorities of occupation and to bring about a rapprochement with the occupying nations; that Smeets was therefore entitled to the protection of the High Commission in virtue of Ordinances Nos. 27 and 70 because of his political activities in the occupied territories during the armistice, and to protection under Ordinance No. 90 because of services rendered these authorities.

The fear expressed by General Allen and by the British High Commissioner that a decision in favor of Herr Smeets would encourage the Separatist movement was justified by articles published later in the "Rheinische Republik." In this organ, Smeets first proclaimed that the Rhineland should be granted autonomy under supervision of the

League of Nations. Later he advocated the establishment of a Rhenish State which should belong to Germany but have no connection with Prussia. In it, all governmental positions should be occupied by Rhinelanders and the administration be free from Prussian officialdom. The German Commissioner from Occupied Territories should be a native of the Rhineland, as well as the members of his staff. A separate budget for the Rhineland should be drawn up and the taxes collected in the Rhineland used in the interests of the Rhineland. The police should be placed under the control of the High Commission in order to prevent oppression of Rhinelanders by Prussian officials. The school system should be subjected to stricter supervision in the interest of international peace.

Most of the above-mentioned proposals were closely allied to policies already adopted by the High Commission. Throughout 1922 the French and Belgian High Commissioners insisted that the majority of officials appointed in the occupied territories be natives of the Rhineland. New appointments to positions as teachers in the schools were rigidly supervised with the view to preventing any instruction which might be prejudicial to the interests of the Allies. It is possible that the views expressed by Smeets were written at the suggestion of the French High Commissioner. In any event, the sentiments expressed in the "Rheninische Republik" coincided with the views of the French press.

Toward the close of 1922, the "Rheinische Republik" became more and more outspoken in its demand for separation of the Rhineland. When the non-Separatist press carried headlines stating that "all Rhinelanders are German and wish to remain Germans," "the Rheinische Republik" reciprocated by stating that the "Rhenish Franks are Rhenish Franks and wish to remain Rhenish Franks." An open campaign was waged against Prussia and Germany and the introduction of a Rhenish currency was strongly advocated. An attempt was made to persuade the population that with the introduction of such a currency, all the evils attending the depression of the mark would immediately disappear. In spite of this campaign, the population, with few exceptions, not only persisted in its loyalty to Prussia but even gave evidence that it was convinced that its salvation depended upon such union. There is little reason to believe that the Rhinelanders will accept separation except through the use of force.

Railway Strikes.

A railway strike broke out in the occupied territories in January, 1922, as had been the case in 1920 and 1921. On December 29th, 1921, information came to the High Commission that the railway employees in the whole of Germany had made demands on the German government for payment of a bonus, and that as a result of the refusal of the German government to comply with these demands, a general strike was imminent. When this information reached the High Commission, the strike had already begun in unoccupied territory and in the British zone of occupation. The High Commission immediately met in extraordinary session and consulted with General Payot, Director of Communications and Supplies for the Armies of Occupation. It decided to inform the representatives of the strikers that the High Commission was willing to receive them and to discuss their grievances with them. From representatives of the German government, also summoned, the High Commission learned that the strike threatened to spread throughout the occupied territories. In view of this situation, the French and Belgian High Commissioners stated that they were ready to declare a state of siege and place the whole of the occupied territories under martial law. General Allen urged the High Commission strongly to consider whether or not there was necessity for such action, because he had received information that the railway management and the strikers, as well, were ready to assure the authorities of occupation that every train designated by them as necessary for their safety and requirements would be run, and that there would be no violence and no disorders.

After long discussion, the High Commission decided to recieve representatives of the railway workers and to inform them that if any railway worker, in violation of Ordinance No. 53, went on strike, he would become amenable to the jurisdiction of the military courts and his services would be liable to requisition, and, in addition, that if a general strike was begun and continued in the occupied territories, the conditions thereby created would compel the High Commission to declare a state of siege.

Ordinance No. 53 of the High Commission forbids the employees of any public service corporation engaged in serving the Armies of Occupation from going on strike until a board of conciliation of the High Commission has attempted to adjust the difficulty. Obviously the railway workmen who went on strike without awaiting the action of such a board of conciliation made themselves liable to prosecution by the military courts. When the workers' representatives were received, they were assured that the High Commission was concerned with their interests and would undertake in accordance with Ordinance 53 to bring about a meeting between them and responsible officials of the German government. The workers replied that they did not see how the High Commission's Ordinance No. 53 could possibly apply, since their employer was the German government which, of course, was outside of the jurisdiction of the High Commission and, therefore, the latter body would be unable to convene the employers or employees before it or before the board of conciliation, as contemplated by the Ordinance. They were informed that the High Commission and the authorities of occupation would not permit a strike at the time. They thereupon replied that the strike had already begun, that they would be unable to control the employees and that any attempt on their part to put an end to it at the time would simply result in the representatives losing their position. However, they gave assurance that all the trains required by the military authorities for their needs would be run.

After this meeting, at which he was present, General Payot proceeded to order a general requisition of all the railway personnel in the Cologne district, This action was based on a misconception of the High Commission's decision, which was to the effect that any workers who went on strike in violation of Ordinance No. 53 might be subject to requisition. General Allen did not think the situation serious enough to warrant the action taken by General Payot and therefore took the position that no order of requisition should be issued or executed in the American area unless the situation made it unavoidable.

As the situation became steadily worse, the High Commission itself decided that a general order of requisition was necessary. It further decided to inform General Payot that it would be expedient to consult the High Commission in future in case a general requisition should appear to be necessary. General Allen concurred with the other Commissioners and caused a limited order of requisition to the put into effect. As a result of the negotiations between the German railway officials and the strikers, the situation soon became normal, but General Payot informed the High Commission that he understood that it was still possible for the strike to be resumed, and that unless the High Commission objected he would not do away with the requisition of the railway personnel in the Cologne district until the excitement had calmed down. The High Commission did not agree with General Payot and informed him that the requisition order must be raised immediately. This order was complied with at once.

Although this strike was not in itself of exceptional importance, the issue of a general requisition order created a very serious situation. Moreover, the declaration of the French and Belgian High Commissioners that they considered it expedient to declare a state of siege was very significant. It had been for a long time the desire of the French military authorities to militarise the railways in the occupied territories, or at least place them under military control. This desire would have been accomplished without doubt if a state of siege had been declared. To a limited extent, indeed, it was accomplished when the railway personnel was requisitioned. Had it not been for General Allen's strong opposition, a state of siege would probably have been declared. Whether or not a state of siege was declared, the railways in any event would have been placed under Allied military control, which control it would have been most difficult to raise. Similarly, it was in large part due to General Allen's insistence that the general requisition order was cancelled.

Information that another strike was threatening on the railways throughout Germany was received later in January. This strike began on February 2nd and effected the larger part of Germany, with the exception of the occupied territories, Upper Silesia and certain parts of

Southern Germany. It did not spread to the occupied territories because the strike leaders were informed that the authorities of occupation would not tolerate any strike movement there. Information was received on February 8th that the issues raised in the strike had been practically settled and that it was to be called off forthwith.

The Commander-in-Chief of the Allied Armies, acting largely through General Payot, made this strike the occasion for insisting that he had unlimited authority over the railways of the occupied territories, and that the High Commission should so decide and publicly proclaim. The Commander-in-Chief of the Allied Armies laid before the High Commission a proposed order, as follows:

The General Officer, Commander-in-Chief, Allied Armies of Occupation, orders:

> "All the personnel of the railways shall continue to discharge their duties and to carry out all instructions given them in his name by the Interallied Railways Commission and the bodies placed under its command.
>
> Any offender against this order will be amenable to the military courts by application of Article 6, Ordinance No. 1, of the High Commission."

Article 6 of Ordinance No. 1 simply enjoins the German authorities and all persons within the occupied territories to obey all orders given by the military authorities in pursuance of their lawful powers.

The order proposed by the Commander-in-Chief assumed complete authority over the railways, not for military purposes only but for all purposes. In spite of this, the High Commission considered an emergency no time to discuss whether this order was within the scope of the lawful power of the military authorities or not, and decided to make no objection. It therefore informed the Commander-in-Chief that it was advisable that he should make known to the military authorities and to the German authorities that his order had received the High Commission's approval. The order was promptly published, but no reference was made to the fact that it had received the approval of the High Commission. This is only one instance of the constant attempts made by the French military authorities to assert their supremacy over the High Commission.

In connection with this same strike, but at a later date, it was proposed that the High Commission enact an ordinance with the following provisions: "Whenever the maintenance, safety or requirements of the armies shall require, the Commander-in-Chief of the Allied Armies, after agreement with the High Commission, will be entitled to requisition all of the personnel of the railways". When this proposition was discussed, the representative of the Commanding General of the French Army on the Rhine contended that under Article 212 of the treaty of peace, the Commander-in-Chief of the Allied Armies had supreme and absolute authority over the railways and that this authority the High Commission could neither control nor limit. Had this view been adopted, the result would have been the militarisation of the railways of the occupied territories. It would have defeated the purpose of the High Commission in large measure and would have charged the military commander in each zone with the control and operation of a vast number of trains even though it appeared the him quite unnecessary.

General Allen felt that the American Army would be seriously concerned if this principle were carried into effect. He was of opinion that it was a very grave matter to assume control of the operation of trains in the American zone or to permit another army to do so. He therefore suggested the following compromise:

> "Whenever by actual or threatened cessation of service on the railways, tramways or waterways in the whole or any part of the occupied territories, military transport is interfered with or is about to be interfered with, or the transportation of the minimum requirements of labor, food and fuel for the civilian population is interrupted or is about to be interrupted, the Interallied Rhineland High Commission may order the requisition of the services of the personnel engaged in the operation of any such service and may order the requisition of the material employed in any such service, in so far as may be necessary to effect military transportation and the transportation necessary to supply the civilian population with its minimum requirements of labor, food and fuel."

After considerable discussion, an ordinance was adopted which was much less sweeping in its scope than the one originally proposed by the French military authorities. Although it did not meet with the full approval of General Allen, he nevertheless in the interests of harmony, gave his assent. This Ordinance -- No. 104 -- provides that the High Commission and not the Commander-in-Chief of the Allied Armies of Occupation shall take the initiative when it is considered necessary to provide for the maintenance, safety or requirements of the Armies and the preservation of public order.

Industrial Strike.

There were many other strikes in the occupied territories during 1922, all of them being settled by boards of conciliation appointed by the German authorities or by the High Commission. In the American area alone there were nineteen, strikes, none of which involved many workmen and all of which, with the exception of the strike of the employees of the Concordia smelting works, were settled without difficulty.

The employees of the Concordia smelting works at Bendorf went on strike on March 20th. This action was a violation of Ordinance No. 53, which provided that employees of public service corporations migh not go on strike until a decision of the disputed points had been given by a board of conciliation appointed by the High Commission. The Concordia works employed about 3,000 workmen, who operated blast furnaces, coke ovens, steel mills and rolling mills. The operation of this plant was of vital importence to Coblenz, as that city was dependent upon the Concordia Huette for its gas supply. Realising the gravity of the situation, the American authorities of occupation induced the employees to return to work, promising immediate action in their case by the High Commission.

On March 21st, at the request of General Allen, the High Commission decided in virtue of Ordinance 53 to constitute a board of conciliation with a representative of the American Department as chairman, and to hear the case. After hearing the views of the employers and the employees, it was obvious that some increase in wages was necessary if the workmen were to remain physically able to perform their labor. The employers were willing to grant a small increase but not to meet the demands of the laborers <u>in toto</u>. The board of conciliation of the High Commission considered the facts presented by both parties and suggested a compromise, which was accepted. The danger of a continuance of the strike was thereby averted and conditions immediately became normal.

Removal From Office of Dr. Momm.

Demonstrations were held by the Socialist and Communist parties throughout Germany on July 4th, in order to show the government that the people insisted upon the enactment of a recently published law which was intended to protect the Republic against attacks of the Right. The immediate reason for the enactment of this law was the assassination of Dr. Rathenau, who was a representative of republicanism. Reactionary elements were responsible for this murder, which caused serious unrest among the more liberal elements of the population. In the course of demonstrations staged in protest in the occupied territories, disorders occurred at Bonn, Cologne and Wiesbaden. The most serious trouble occured at Wiesbaden, where the police authorities appeared to have made no proper provisions for controlling the crowd and exercised no constraint over the policemen on duty. When certain members of the crowd became violent, the police intervened and, acting with very poor judgment and apparently without provocation, opened fire, killing one person and wounding a number of others. The French authorities of occupation immediately preferred charges against Dr. Momm, Regierungspraesident of the Government District of Wiesbaden, for neglect of duty and failure to take the necessary measures to insure public order. When Regierungspraesident of the District of Treves, he had been highly thought of by the officials of the

American Army there, and, during the early days of the American occupation, rendered valuable service in adapting German laws to the needs of the military occupation. He was suspended from office by the French authorities in 1920 because he was suspected of having participated in the arrest and abduction of Dr. Dorten, the erstwhile leader of the Separatist movement and President of the so-called "Republic". Dr. Momm was later cleared of these charges and reinstated in office.

The French General commanding the local area of Wiesbaden urgently requested the removal of Dr. Momm, asserting that the charges against him were simply a culmination of many disagreements and difficulties. The French High Commissioner also informed the High Commission of a number of disagreeable events that the French authorities had experienced in their administration of the area under Dr. Momm's jurisdiction.

The incidents which formed the basis of the charges did not appear of a character sufficiently grave to justify an effort to remove the governor of a large area. Although two men were killed and a number were wounded, the collision with the police was clearly unpremeditated and resulted from mob passion. Germans only were properly concerned in the matter. No allied soldier was injured or threatened, nor was the safety of the army in any way endangered. It was quite evident that the French authorities were influenced by political considerations and decided to take advantage of the incidents at Wiesbaden to gain the good will and support of the populace by demanding the removal of an official whose removal they thought would be most agreeable to the people in general. The facts proved that they were very much mistaken. Although the demonstrations were not friendly to Dr. Momm, because he belonged to the party of the Right against which the demonstration was organised, there was no question in the minds of the people when they were to decide whether they would align themselves with their own officials or with the French authorities, and they immediately united in urgent petitions to the High Commission defending Dr. Momm and protesting against any action that might be taken towards his removal.

General Allen and the Belgian and British High Commissioners expressed the view that the evidence submitted did not warrant Dr. Momm's removal. The French High Commissioner disagreed, and moreover asserted that it would be impossible for the French authorities of occupation to cooperate with Dr. Momm. He asserted that in view of the publicity given the charges against the Regierungspraesident and in view of the protests made by German officials, organisations and the press, the loss of prestige to the French authorities, should the High Commission fail to sustain the charges, would be so great that further administration of the area would be most embarrassing. Although General Allen and the Belgian and British High Commissioners were not convinced, they realised the impossible situation in which the French High Commissioner and the military authorities found themselves as a result of the Wiesbaden affair and the necessity of taking some action to alleviate this situation. It was therefore decided to point out to the German Commissioner the unfortunate situation at

Wiesbaden and to ask him to request his government to transfer Dr. Momm and replace him with an official more acceptable to the occupying authorities.

After the lapse of a time limit set by the High Commission, the German Commissioner informed the High Commission that the Prussian government appreciated the proposal of the High Commission, which was made in a spirit of accommodation; the Prussian government was ready to make use of the earliest opportunity to transfer Dr. Momm to another post, but would not be able to effect this transfer until sufficient time had elapsed to prevent the German people from connecting it with the incidents of July 4th. The High Commission then proposed to the German Commissioner that no action would be taken against Dr. Momm for a period of three months provided the Prussian government give him a leave of absence for that length of time. This proposal was made with the idea that the Prussian government would be able to effect Dr. Momm's transfer within three months. The Prussian government's reply to this suggestion was that it was willing to transfer Dr. Momm to another post and would take the earliest practicable occasion to do so, but that it was unable at the time to grant him a leave of absence. The High Commission did not consider this reply satisfactory and decided to dismiss Dr. Momm from office.

This whole incident was extremely unfortunate. It had been obvious for a long time that the French authorities wished to get rid of Dr. Momm and it was evident that they were making use of the first opportunity to do so. The action of the French authorities in informing the public that Dr. Momm would be expelled placed the High Commission in a delicate position. The High Commissioners, for the sake of harmony and also in order to save the position of the French, supported the French High Commissioner; but they were not at all convinced that his action was correct.

Attempt to Encroach on General Allen's Authority.

When the majority of the American troops were withdrawn in 1922, the Commanding General of the French Army on the Rhine agreed that the French troops who replaced the American troops in the American area should be under General Allen's orders. In spite of this fact, General Degoutte, Commanding General of the French Army on the Rhine, asked the High Commission to authorise the establishment of an office of the Reichsvermoegensverwaltung for the French troops in the American zone. The Reichsvermoegensverwaltung is an organisation established by the German government in the occupied territories to provide for the needs of the armies and especially to comply with the demands of the authorities of occupation with respect to billets, quarters and fiscal matters as provided in the Rhineland agreement. One branch of this organisation was attached to Headquarters of the Commanding General of each of the occupying armies and attended to all matters concerning the troops in the zone occupied by that army, under the supervision and direction of the Army Commander. In his letter to the High Commission, General Degoutte stated that he asked for the establishment of such a branch, whose special duty it would be to deal with the requirements of the French troops, in order to facilitate financial control of the work of the Reichsvermoegensverwaltung, and because he believed that if the needs of both the American and French troops were managed by the same office, difficulties, delays and mistakes might arise. He therefore suggested the creation of an organisation which should act under the Commander of the French troops in the American zone and under the branch of the Reichsvermoegensverwaltung in the French area, instead of that in the American area.

When this question was brought to the attention of General Allen, he informed General Degoutte that his proposal was directly at variance with the signed agreement made at the time of the entry of the French troops into the American zone, and that the establishment of any organisation according to the plans of General Degoutte, independent of the control and direction of General Allen, would affect the prestige of American representation in the occupied territories and would therefore not be allowed. General Degoutte was further informed that the American administrative services were looking after all the needs of the French troops in a satisfactory manner and that there had been no complaint in regard to these matters. The High Commission was of opinion that the whole question was not within its competency and that it was one which concerned the Army Commanders alone. The result of negotiations between the American and French commands was that no separate branch of the Reichsvermoegensverwaltung was established.

the French authorities to separate the French troops in the American zone from American administration. It was only natural that there should be a certain unwillingness to have French soldiers under the command of an American general, but the existence of a signed agreement should have prevented such steps. An attempt was made at a later date to have assessment commissions in the French zone assess all claims against French troops in the American area. This was in violation of a decision of the High Commission, whereby claims against troops of another nationality than that of the troops in the particular zone should come before a commission situated in the zone where the requisition had been made or the damage had been caused. When this question was brought to the attention of General Degoutte, he realised that his subordinates were in the wrong and acknowledged the competence of the assessment commission in the American area. This and similar questions were obviously not within the competence of the High Commission and were therefore left to the military authorities to adjust between themselves.

Western European Time in Rhineland.

As was the case in October 1920 and 1921, it was proposed that western European time be introduced in the occupied territories in October, 1922. It is the custom in France and Belgium to change from summer to winter time in October, thus bringing about a difference of one hour between western and central European time. Since the beginning of the occupation, railway time in the occupied territories had been changed each year, to conform with western European time. In the beginning, this change was made at the request of the military authorities who stated that should it be necessary to mobilise and send troops from France and Belgium to Germany in case of disorder or an attack on the troops of occupation, a difference of time migh prove disastrous, or might cause serious difficulties.

No objection was raised to this system until 1920, when the British High Commissioner, under instructions from his government, pointed out that since the treaty of peace had come into effect and conditions in Germany were practically normal, there was no reason to expect any acts of aggression, and that there was therefore no occasion for the public to be subjected to the serious inconvenience arising from a difference of time between occupied and unoccupied Germany. However, since the Interallied Military Committee of Versailles had expressed the opinion that it was advisable that the same railway time should be used in the occupied territories as in Belgium and in France, the High Commission decided that the change of time should take place. The same question arose in 1921, and was settled in the same way, and for the same reason.[1]

This question was also brought up in September, 1922. The Commanding General of the Allied Armies of Occupation requested that the usual change in time be made in order to

---

[1] A. R. in O. G. 1920-21 Vol. I., page 174

to the Occupied territories in the event of serious disturbances. This recommendation was supported by the French and Belgian High Commissioners, although the latter stated that he was personally not in favor of such a change but had been instructed to vote in favor of it. The British High Commissioner expressed the view of his government to the effect that military requirements were not of sufficient importance and the general situation not of sufficient gravity to justify the adoption of western European time when due regard was paid to the interference which this measure would entail to civilian life in the Rhineland. General Allen expressed the view that, from a military viewpoint, it would be convenient, but not necessary, to change the time in the occupied territories, and that the military convenience to be served would not justify the inconvenience to the civilian population which would result from the change. One of the principal reasons which led General Allen and the British High Commissioner to object to this measure was that they had been informed that this change of time was likely to be permanent, as it was understood that neither France nor Belgium intended to reintroduce summer time. General Allen was moreover of opinion that the measure was simply part of a policy tending to separate the occupied territories from the rest of Germany. His belief in this respect was strengthened by the fact that although the British High Commissioner suggested a number of solutions in order to reach a compromise, the French and Belgian High Commission, by a majority vote, decided that western European time should be adopt-

This was the first time that the High Commission had taken a decision by a majority vote instead of unanimously.

This fact is an indication of the constantly decreasing influence of Great Britain in the High Commission. This decrease in influence became manifest upon the departure of Sir Harold Stuart[1] and his replacement by Mr. Robertson.[2] Lord Kilmarnock,[3] the successor of Mr. Robertson, enjoyed much less prestige than had his predecessor. As a result of his failure to offer strong resistance to measures which his advisers did not consider expedient, the British Department adopted a policy of non-resistance. It was therefore possible for the French and Belgian High Commissioners to obtain the adherence of the British Department to many steps which did not appear justified by the Rhineland agreement. Another result was that the French and Belgian High Commissioners no longer hesitated to adopt a decision by a majority vote.

Policy of High Commission with Respect to German Officials.

In October, 1922, a French military court tried eight German nationals who were accused of having communicated to a German Intelligence service in non-occupied Germany military information regarding tactics of the French Army, distribution of troops in the occupied territories, manoeuvres, training camps and garrisons. The sentences imposed ranged from two to twenty years' imprisonment. Before the trial began, the

---

[1] A. R. in O. G., 1920-21 Vol. p., pp. and p.
[2] Ibid, p. 266. [3] Ibid, p. 267.

German Commissioner protested against the arrest of a number of the accused and stated that it was difficult to see how they had rendered themselves liable to punishment, because they had communicated facts which were not military secrete and which were accessible to almost everybody. He maintained that the communication of information concerning the armies of occupation was punishable only when it dealt with matters that were secret; the information which the accused were alleged to have furnished was contained in the Allied newspapers published in the occupied territories and accessible to all of the inhabitants of both occupied and unoccupied Germany. The High Commission took no action other than to forward the protest to the French military authorities. An explanation of the severity of the sentences imposed was perhaps to be found in the fact that the French military authorities wished to inspire fear in the agents of the German intelligence service.

In December, 1922, a French military court tried Dr. Frange, a high German official whose work mainly concerned matters regarding the occupation. Dr. Frange was sentenced to imprisonment for five years because he was alleged to have collected information prejudicial to the safety of the armies of occupation and to have held communication with powers other than those participating in the occupation. It was doubtless true that Dr. Frange assisted in the compilation of material for a German campaign of propaganda against the French authorities, but it was difficult to conceive how his work could be considered an act of espionage.

Possibly the French military authorities were unusually severe in this case because they wished to frighten other German officials so they would not act as agents of the German government in the collection of material destined to be used against the authorities of occupation. The action taken in this case and those immediately before it was probably intended to intimidate officials to such an extent that they would refrain from reporting instances which might tend to show that the authorities of occupation had abused their power.

The High Commission noticed in 1921 that the number of non-resident officials appointed to positions in the Rhineland was constantly increasing. The High Commission was of opinion that these appointments were made by the German authorities in order to maintain Prussian influence through reactionary Prussian officials. With this in mind, the authorities of occupation vetoed the appointment of a large number of officials, especially school teachers, from East Prussia and Silesia. In 1920-21, twenty-three officials were removed from office and the appointments of twenty-eight other officials were vetoed.[1] The High Commission became even stricter in 1922 and vetoed the appointments of 174 officials. On the other hand, only one official of importance -- Dr. Momm -- was removed from office.

This great increase in the number of vetoes may be ascribed to a desire on the part of the authorities of occupation to alleviate the lot of Rhinelanders by requiring the

---

[1] A.R. in O.G. 1920-21, Vol. 1., pp. 222.

German authorities to appoint mostly natives of the Rhineland to office. It may also be ascribed to a desire to further the Separatist movement by requiring the appointment of residents of the Rhineland who might be expected to be at least non-sympathetic to Prussia. In any event, the High Commission's policy during 1922 was to require that at least 50% of the officials in the Rhineland be residents of the Rhineland and especially not residents of Prussia.

Schools.

Many of the appointments that were vetoed were those of school teachers. These vetoes were imposed because the High Commission had been informed that the percentage of school teachers and instructors appointed to positions in the occupied territories and who were not natives thereof was constantly increasing. The High Commission had also been informed that the provisions of the German constitution providing for education in the spirit of reconciliation of nations were not being complied with. Teaching in high schools and universities was said to be reactionary and monarchial wherever socialist influence was not too strong. This inclined the High Commission to believe that teachers were being brought from unoccupied Germany to the occupied territories for the purpose of checking the spirit of reconciliation and in order to keep alive the feeling of hatred toward the Allies. With this in mind, the High Commission appointed a special committee to consider what measures should be taken to confer with German educators in order to ascertain what might be done to rectify the situation.

From the report submitted by this committee of inquiry it appeared that while the teaching in schools in occupied territory was in general objective, anti-Entente propaganda was being carried on, especially in high schools and universities. With respect to the introduction of non-Rhenish teachers, it was of opinion that the instructions recently given on this point by the Prussian government, viz., that nominations to positions in the occupied territories, would no longer be made at Berlin, with the exception of a few communal teachers, should be taken into consideration. The committee therefore proposed to the High Commission that the number of vetoes of nominations of members of the teaching profession originating in non-occupied Germany be reduced. In spite of this report, there was no decrease in the number of vetoes during the latter part of 1922.

Passau-Ingolstadt Incidents.

In December, 1922, the conference of Ambassadors decided to impose a fine of 1,000,000 gold marks upon the cities of Passau and Ingolstadt in Bavaria because of certain attacks upon representatives of the Interallied Military Commission of Control. On December 8th, the High Commission was informed by the Conference of Ambassadors that if the fine were not paid by December 10th, it should insure the collection or seizure of the equivalent of that sum from the revenues derived by Bavaria from the Palatinate. Upon receipt of these instructions, the High Commission prepared a draft ordinance providing that all monies received from the collection in the

Palatinate of duties and taxes, or from the administration of State property and, in particular, customs receipts, proceeds of taxes and sums accruing from the administration of State forests, be sequestered and placed to a special account. The departments of finance, customs, forests and, in general, all the personnel in the Palatinate engaged in the collection of taxes and the administration of State property were made subject to the authority of the High Commission. At the same time, the High Commission prepared another draft ordinance providing for the constitution of an executive committee to be charged with the execution of the measures outlined above. These measures were to come into force on December 11th if Germany had not paid the fine of 1,000,000 gold marks. Inasmuch as the fine was paid on the prescribed date, the High Commission did not have to put its draft ordinances into force. The German reply to the demands of the Conference of Ambassadors was not completely satisfactory, but no further measures were taken at the time.

The ultimatum delivered to Germany appeared to have been an adroit move intended to create discord between the Palatinate and Bavaria and between Bavaria and the rest of Germany. This was to be assumed from the fact that it was Bavaria and not the national government which was called upon to pay the fine and tender the apologies. The provision that a levy would be made upon the resources of the Palatinate in case of non-payment seemed to be directed toward arousing the population of the Palatinate against the Bavarian authorities because the latter preferred to sacrifice the interests of the people in the occupied territory rather than allow Bavarian pride to be humiliated. On the other hand, the direction of the note to Bavaria and not to Germany as a whole was probably intended to lead to the belief that Bavaria was to be regarded as a separate State and not as a part of the German republic. The German government realised this, because in its reply it stated that the various German States in their relations with foreign countries were represented by the central government and that the German national constitution did not permit correspondence by the respective German State governments with foreign powers. The German government also recognized the danger which threatened the occupied territories if satisfaction were not given the Allies, because the German note stated that although the German government did not recognize the Allied demands as just, it nevertheless would comply with them in order to protect the occupied territories from the impending punitive measure.

The French press frankly stated that the expressed intention of the Allies to seize pledges in the occupied territories in case of non-payment was the first step toward carrying out the policy of making German debtors pay direct to their Allied creditors. It was further stated that the first of the productive pledges at the disposal of the Allies was easily available in the occupied territories. From events which occurred at a later date in 1923 -- the occupation of the Ruhr-- it appears that the intention to exploit the Palatinate was simply an indication of a much more extensive plan.

Action of High Commission with Respect to German Press

The attitude of the High Commission with respect to the German press became increasingly severe during the year 1922. It was found necessary to exclude, suspend or prohibit forty-two publications in 1920. In 1921, such action was taken with respect to fifty-five publications.[1] In 1922, no fewer than ninty-four publications were excluded, suspended or prohibited. The reason for this extraordinary increase was in large part the fact that the High Commission took action against publications not only because they attacked the authorities of occupation but also because they published articles directed against the governments participating in the occupation. For instance, the "Koelnische Zeitung", one of the most important newspapers in Germany, was suspended because it published a series of articles concerning the question of responsibility for the war. Most of the requests for action against publications came from the French and Belgian authorities. Very few came from the British and only one from the Americans.

New Legislation Enacted by High Commission.

During 1922 the High Commission continued its policy of providing an ordinance for every situation that might arise, and also made numerous changes in the legislation previously enacted by it. Thus, the number of ordinances was increased from 102 to 127; the number of instructions was increased from 17 to 19.

---

[1] A.R. in O.G., 1920-21, Vol. 1., page 227.

The most important ordinances enacted by the High Commission in 1922 were the following:

Ordinance No. 113, regarding dissolution of customs managing board of I.A.R.H.C.

Ordinance No. 116, regarding application in occupied territories of German code of criminal procedure.

Ordinance No. 118, regarding exemption of members and personnel of High Commission and of armies of occupation from taxes.

Ordinance No. 120, regarding, first, transfer from occupied territories of persons arrested by German authorities and claiming benefit of Ordinances Nos. 27, 70 and 90; secondly, powers in occupied territories, of German judicial and police authorities of non-occupied Germany.[1]

Departure of M. Cattoir.[2]

The only important change that took place among the important officials of the High Commission in 1922 was caused by the departure of Belgian Deputy High Commissioner Cattoir, who was summoned to fill an important post in the Belgian Ministry of the Interior. M. Cattoir was a most earnest collaborator in the work of the High Commission; but he almost always supported the French position in the fulfilment of his duties. He was succeeded by Count de Liedekerke,[3] former legal adviser of the Belgian Department.

ACTIVITIES OF INTERALLIED RHINELAND HIGH COMMISSION IN 1923.

Occupation of Ruhr.

---

[1] Copies of Ordinances and Instructions issued by High Commission in 1922 are contained in a separate appendix. [Not reproduced in mimeographed edition]

[2] A.R. in O.G., 1920-21, Vol. I, p. 263

[3] A.R. in O.G., 1920-21, Vol. I, p. 265

As a result of Germany's failure to comply with the provisions of the treaty of peace and the demands of the Reparations Commission, French and Belgian troops began the occupation of the Ruhr Valley on January 12th, 1923. The French Commander-in-Chief issued a proclamation in which he stated that the French and Belgian governments had decided, in accordance with the treaty of Versailles, to send into the Ruhr a control commission consisting of engineers and invested with power to oblige Germany to respect the obligation it had contracted. The proclamation continued :

> "The object pursued by the French government is absolutely pacific.
> "In sending this commission it declares that it has no intention of undertaking a military operation or a political occupation.
> "Only such troops as are necessary to protect the commission and guarantee the execution of its duties are being sent into the Ruhr."

At a meeting of the High Commission on January 16th, the French High Commissioner reported that he had received instructions from his government directing him to propose to the High Commission the application in the occupied territories, through ordinance of the High Commission, of the measures proposed by the French government as economic reprisals. He stated that in accordance with these instructions, he had prepared three ordinances which he would ask the High Commission to consider immediately. The Belgian High Commissioner reported that his instructions were of like tenor. The British High Commissioner reported that his instructions were to the effect that he was not to participate in any action or discussion concerning measures of the kind proposed by the French High Commissioner. The British High Commissioner received a dispatch from his government during the meeting. This he announced, so far modified his instructions as to permit him to participate in meetings concerning these measures but would not permit him to vote on them. Apparently this modification of instructions was brought about in order that the meeting might be properly constituted, even though the British High Commissioner opposed the measures or refrained from voting.

Thereupon the High Commission considered and adopted three ordinances, viz., 1. providing for seizure of the coal tax collected in occupied territories, 2. seizure of all customs receipts in occupied territories and impounding of funds accumulated from collection of customs wherever such funds were held in occupied territories : (this ordinance did not contemplate taking over the customs operations but did contemplate that Allied agents would be stationed at all points where customs were collected within the occupied territories in order that the monies collected might be seized at the source). 3. seizure of all proceeds resulting from exploitation of forests in occupied territories.

At the conclusion of the meeting, the High Commission decided upon a communication to be inserted in the Allied press at the request of the British High Commissioner, who expressed the hope that the communication would be copied in the German press. The communication was as follows:

> "At a sitting held this morning, the High Commission, in consideration of instructions received by certain of the High Commissioners from their governments, promulgated three ordinances regarding :

(1) seizure of coal tax in occupied territories;

(2) seizure of customs receipts in occupied territories;

(3) seizure of certain pledges in connection with management and exploitation of forest resources in occupied territories.

"Acting upon instructions received from his government, the British High Commissioner was present at the sitting but refrained from voting. The Italian representative was also present."

On the occasion of the occupation of the Ruhr, a representative of the German Commissioner called upon the American Department and invited its attention to the protest of the Russian Soviet Government against this Occupation. He placed special stress on the following passage of the protest :

"In this fateful day, the Russia of the workers and peasants raises its voice again in passionate protest against the insane policy of imperialistic France and its Allies."

After the reading, he exclaimed : "If no powerful western government help us, is it not clear that we shall be thrown into the arms of the East ?" The German representative was asked whether Germany was trying to use Bolshevism as a bogey to frighten France. It was intimated to him that the measure would not have a chance to succeed unless Bolshevism became real, when Germany would be undone. It was suggested that suicide as a protest was not practised in the west. The German deputy replied that the destruction of Germany would result in the destruction of France and that Germans were fast approaching a state of mind in which they should suffer the one in order to bring about the other.

Resistance of German Officials.

The ordinances enacted by the High Commission at its sitting of January 16th resulted in open defiance by German officials, high and low, throughout the territories occupied pursuant to the treaty as well as those newly occupied. This attitude was assumed on orders from the German State and national governments, which orders appeared to have met with ready, willing and general obedience by all officials of whatever grade. Many instructions were issued by the governments concerned, of which the following from the State governments of Prussia, Bavaria, Hesse and Oldenburg were typical;

> "The action of the French and Belgian governments in the Ruhr District constitutes a grave violation of international law and of the Versailles treaty. By reason of this, all orders and instructions issued to German officials in the execution of this action are legally of no effect. The national government and the government of Prussia, Bavaria, Hesse and Oldenburg therefore direct that orders of the occupying powers are not to be followed but that the orders of our own government are to be adhered to exclusively. This applies to all officials of the territory formerly occupied with reference to all measures in conflict with the terms of the Rhineland agreement.
> (Signatures) Cuno, Braan, v. Knilling, Ulrich, Tautzen."

Similar orders and instructions were issued by and to the various German services concerned in the application of the newly enacted ordinance.

In consequence of their refusal to obey these ordinances, the High Commission ordered many officials to be deported from the occupied territories. In a number of cases the order included the families of the officials deported. At a single sitting, the High Commission decided to deport no fewer than forty-eight higher officials, including governors of large districts, and important customs, finance, forestry and municipal officials. At first, the High Commission saw fit to deport only officials who had definitely refused to obey specific orders, but a number of others invited dismissal and deportation by defiant manifestoes announcing in advance their intention to disobey such orders. A typical example was one signed by the highest executive officer in the Rhine Province, whose office is at Coblenz, together with all the high functionaries representing the province, district and municipality. It was as follows :

> "Under date of January 20th, 1923, the German and Prussian governments issued instructions to the effect that all orders and ordinances decreed as a result of the action of the French and Belgian governments in the Ruhr basin should not be executed by German officials, as being without legal effect; that on the other hand, the said officials must carry out the instructions of their own governments exclusively and that this applied equally to the officials of the old occupied areas in respect of all measures contradictory to the stipulations of the Rhineland agreement."

It was to be expected that action would be taken against those who signed this communication. The first to suffer was Dr. Fuchs, Ober-Praesident of the Rhine Province, who was deported on February 2nd.[1]

Although most of the officials immediately deported were from the French and Belgian zones, one of the first was the chief of the Finance Office at Cologne, in the British zone. Following this action, and evidently in accordance with instructions from his government, the British commander-in-Chief announced that for the time being, or until new instruc-

---

[1] See page 136.

tions were received from the British government, no further deportations or arrests of German officials would be made in the British zone.

When the orders impounding the accumulated customs funds in certain banks were issued, the French and Belgian authorities reported that the Reichsbank in certain cities, among them Duesseldorf, Aix-la-Chapelle and Dortmund, had immediately begun to move their assets to non-occupied Germany. This removal was then forbidden by military orders and a number of the Reichsbanks were seized. They were subsequently released and a military order impounding accumulated customs funds was substituted; but in the meantime, and as a protest, all the banks in the towns where the seizures had taken place were closed. The banks resumed business when the order was modified.

Application of Ordinance 49 in American zone.

As soon as General Allen had turned over command of the American zone to the French authorities, the High Commission adopted Ordinance 139, applying the provisions of Ordinance 49, concerning billets and quarters, to the Landkreis and Stadtkreis, Coblenz. When Ordinance 49 was adopted in 1920, General Allen requested the High Commission not to have it apply to the Landkreis and Stadtkreis, Coblenz, because the congested conditions in Coblenz required special regulations. Moreover, General Allen could not admit that in the matter of billets in the American zone, the High Commission should have rights superior to his own. The High Commission complied with General Allen's desire, and Ordinance 49 was not applied in Coblenz.[1]

It provided that the allotment of billeting and quartering accommodation as between the High Commission and the military authorities in the urban and rural Kreise of Coblenz should be managed by a billeting commission consisting of a representative of each of the Departments of the High Commission and of an equal number of representatives of the military authorities. The President of the High Commission was made chairman of this Commission.

Deportation of Officials.

Since German officials consistently refused to obey its new ordinances, the High Commission proceeded to expel them in ever increasing numbers. Thus, at the meeting held on February 1st, 1923, it decided to deport fifty-four more higher officials. The number of deportation from the newly occupied territory was even greater, as the officials there were even more defiant in refusing to obey the orders of the authorities of occupation.

Attitude of British.

The attitude of the British with respect to all these measures is best illustrated by the British reservation inserted in the minutes of the High Commission in which the action of the High Commission with regard to the measures of reprisal was recorded. It was as follows :

"The British High Commissioner stated that

---

[1] A.P. in O. . 1920-21, Vol. I., page 161.

in accordance with instructions received from his government, he would refrain from taking part in this decision and disclaimed all responsibility for it on behalf of his government and himself."

As the British government refused to associate with France and Belgium in the measures of reprisal, the position of the British authorities of occupation became increasingly difficult. The following incidents illustrate this :

The offices of the German Coal Administration in the occupied territories were at Cologne, in the British zone. The German officials of the Coal Administration at Cologne felt that they were more or less immune from molestation by the French and Belgian authorities because of the policy of the British government not to allow any arrests to be made or punitive measures to be taken without its consent. The result was that the German officials refused to help the Allied officials in the collection of the coal tax or in the distribution of coal. In spite of unofficial intervention by the British, the Germans refused to cooperate in the collection of the coal tax, although consenting to assist in the allocation of supplies. The French and Belgians became very indignant and wished to take strong action, but the British would not permit this in view of the instructions of their government. The British were in an impossible position, because if they continued to refuse to let the French and Belgians take action, the Germans would construe that attitude as being in support of Germany and the French and Belgians would to that extent be unable to apply their sanctions, whereas if the British determined to permit sanctions to be applied, they would either have to violate their own government's orders and apply the sanctions through their own courts or permit the French and Belgians to establish courts in the British zone.

One of the most influential of German newspapers, the "Koelnische Zeitung", printed at Cologne in the British zone, published an article calling upon all German officials and the German people to resist the actions of the French and Belgians. The Belgian High Commissioner took exception to this article and informed the High Commission that the German officials in the British zone felt that they were in a stronghold which could not be attacked by the French and Belgians because they enjoyed the protection of the British. The French and Belgian High Commissioners stated that this position was impossible and that the resistance would not be tolerated; that the British must either permit the French or Belgians to down their resistance or do it themselves. Either action would have implicated the British contrary to direct instructions received from their government. In this case the French and Belgian High Commissioners decided that the "Koelnische Zeitung" should be forbidden in the French and Belgian zones and that no action should be taken in the British zone. The British High Commissioner was requested to intervene unofficially with the editor of the "Koelnische Zeitung" and persuade him not to publish any articles encouraging resistance.

From these instances it became increasingly evident that the position of the British authorities of occupation was becoming embarrassing. It became evident that the British would have either to collaborate with the French and Belgians or withdraw completely from the Rhineland. The fact that trains in the British zone were operated by German employees while a general railway strike prevailed throughout the occupied territories further served to demonstrate that the Germans were deliberately trying to create discord between Great Britian and her Allies. Although the British authorities of occupation realized this, they did not know what to do because their instructions prevented them from taking action in any direction.

Assumption of Control of Import and Export Licenses by High Commission.

In March, 1921, as a part of the sanctions then applied, the High Commission took over the licenses for all imports and exports crossing the German frontiers into France, Luxemburg and Belgium.[1] After October, 1921, when the sanctions were supposed to have been removed, the High Commission still maintained supervision and control over import and export licenses. The offices of this service were at Bad Ems. Following the latest sanctions, the German government issued an order to the effect that all licenses issued by Bad Ems were void and were to be given no effect. The High Commission thereupon directed the Bad Ems service to pay no attention to the German order. Evidently acting upon instructions from the Berlin government, the German officials and employees of the Bad Ems service immediately went on strike and destroyed their records. The High Commission therefore decided to dismiss all the German customs officials in the occupied territories and to reestablish all these services with Allied personnel.

[1] A.R. in O.G. 1920-21, Vol. I., pp. 180-2.

Immunity for German Officials Obeying Instructions of High Commission.

The High Commission was of opinion that a number of German officials might obey its orders in spite of instructions received from Berlin, provided they were assured of immunity against prosecution by the German authorities. It therefore decided to enact Ordinance 140, granting immunity from prosecution by the German government to all German officials who took service with the High Commission in the department of customs.

Occupation of Further Territory in Baden.

As a result of the statement by the German government that, on account of shortage of coal, the Germain railways would no longer be able to operate through Germany the Paris-Orient express, running from Paris via Prague to Warsaw, the French government ordered the High Commission to take sanctions against the German railways. In compliance with its instructions, the High Commission ordered the occupation of the railway junction-points of Offenburg and Appenweier in Baden, opposite the Kehl bridgehead, and directed that all trains should be stopped indefinitely at these points, except the Holland-Switzerland express from Amsterdam to Bale. There was no contention made that the Rhineland agreement furnished authority for such action. This measure, together with the measures referred to above, indicated that the High Commission was being used more as an instrument for applying new reprisals than as the supreme representative of the Allied and Associated governments in the execution of the Rhineland agreement. As a matter of fact, upon the occupation of the Ruhr and the refusal of the British to participate in this occupation, the High Commission became merely the agent of the French and Belgian governments. The British High Commissioner, acting upon instructions from his government, became a mere observer who was present at the meetings but withheld his opinion in all important decisions.

Establishment of French Kreis Delegate in Coblenz District.

On January 27th, 1923, the American representatives in the several German *Kreise* turned over their duties to their French successors. There was a great difference between the number of American personnel and that of French personnel considered necessary to administer these districts. The American district delegate who had supervision of all the *Kreise* was served by four clerks and one chauffeur. His French successor had one assistant and twenty-two clerks. The district of Coblenz was divided into a rural and an urban *Kreis*. The American organization administering both urban and rural *Kreise* consisted of one representative, one assistant, one chauffeur and one clerk. The French organization consisted of one representative, one assistant, one chauffeur, three clerks and three soldiers for the urban *Kreis* and exactly the same for the rural *Kreis*, making a total of eighteen, as against four Americans. As administered by the Americans, the *Kreise* of Mayen, Cochem, Montabaur and Neuwied had three men each. The *Kreise* of Ahrweiler and Adenau were administered together by three Americans. The French had nine men in each *Kreis* and began the administration of Ahrweiler and Adenau separately. In all, the French organization which succeeded the American consisted of ninety-six persons as against the American twenty-five; furthermore, the French stated that this was a mere tentative organization and would be increased later.

Separate Currency for Occupied Territories.

The question of a new currency for the Rhineland was discussed by the High Commission in the latter part of 1922. This question assumed more practical importance in 1923, because the German government then refused to furnish money to pay either the French and Belgian army or civilians. In the Belgian area, resort was had to the requisitioning of money in part from private banking institutions in order to meet the military payroll. In February, 1922, it was proposed to establish a new currency for the Rhineland, to be issued under the authority of the High Commission. This currency was to be based on the German State-owned mines in the occupied territories and on the general security provided by Article 248 of the treaty of peace, which states that reparations shall be a first charge upon all the assets and revenues of Germany and its constituent States. The basis for this currency seemed to be rather insecure because there was only one productive State-owned mine in the occupied territories. Moreover, the general pledge provided by Article 248 of the treaty of peace was so indefinite as to be incapable of enforcement against resistance except through the application of military power. Furthermore, both of these so-called securities were already pledged for the whole reparations debt and if they were realised upon would naturally be applied in payment of reparations. There is little doubt that a separate currency for the Rhineland will be introduced if the occupation continues, but upon what it will rest is difficult to predict.

Special Ordinances Issued by High Commission to Enforce Measures of Reprisal.

In addition to the ordinances originally enacted by the High Commission to enforce the measures of reprisal adopted by the French and Belgian governments, the High Commission found it necessary to issue further ones, to wit :

Ordinance No. 135, establishes customs managing committee.
import and export license managing committee,
forest managing committee,
special coal and mines committee and
special accounting committee;

Ordinance No. 136, regulating collection of customs duties at frontiers of occupied territories and providing for delivery of licenses for exporting and importing goods from and to occupied territories;

Ordinance No. 137, regarding distribution of coal in occupied territories;

Ordinance No. 138, regarding production of coal in occupied territories;

Ordinance No. 140, regarding protection of officials and employees and private individuals obeying special ordinances of High Commission.

Withdrawal of American Representation on Interallied Rhineland High Commission.

On Tuesday, February 6th, 1923, General Allen informed the President of the High Commission that in view of the withdrawal of the American Forces in Germany, his government felt that the circumstances which warranted it in maintaining an observer with the High Commission no longer existed. General Allen further expressed his government's sincere appreciation of the many courtesies which the High Commission and its members had shown to the government of the United States and to its observer.[1]

M. Tirard, President of the High Commission thereupon proceeded to deliver a farewell address for the High Commission. He expressed his appreciation of the success achieved by General Allen in dealing with the delicate situation which arose as the result of the failure of the United States to ratify the threaty of Versailles. He said that it was chiefly due to General Allen's courtesy and diplomacy that for three years, in the most varying circumstances, no difficulties had arisen from the peculiar situation in the American zone. He further remarked that everybody in the High Commission had been impressed by the highmindedness and perfect impartiality of General Allen's judgment. He expressed sincere regret that this was the last occasion upon which he would have the opportunity of saluting the representative of the United States of America. He declared that he would always associate the remembrance of General Allen with

[1]See page 219.

the same magnificant spirit which, in the hour of danger, united the troops of his country and those of the Allies in their struggle for like principles of justice and liberty. He also extended his thanks to the whole of the American delegation and particularly to Colonel Stone[1] and Major Manton Davis[2], who had lent their valuable collaboration in the work of the High Commission and whose labors were incorporated in its ordinances and decisions.

Baron Rolin-Jacquemyns, the Belgian High Commissioner, associated himself with what was said by Monsieur Tirard. He also stated that General Allen was better able than any other to express the feeling of union which had never ceased to reign among the Allies.

Lord Kilmarnock, British High Commissioner, asserted that he had always appreciated the tact and courtesy which General Allen had constantly exhibited and the spirit of high-minded impartiality with which he had treated the many delicate questions that had arisen. It had been a great pleasure to have him present at the meetings of the High Commission and his high qualities had always been appreciated. He expressed his extreme regret that this was the last time that General Allen would be present at a meeting of the High Commission.

To these addresses General Allen replied that it was very difficult for him to find words with which to express his personal gratitude for the great courtesy, consideration and unfailing generosity which had at all times been extended him

---

[1] A.R. in O.G., 1920-21, Vol. I, p. 257.
[2] A.R. in O.G., 1920-21, Vol. I, p. 259.

and his collaborators. He assured the High Commission that he had at all times worked in common counsel with a view to solving the many difficult and delicate problems regarding the government of the occupied territories and the relations of the military forces to the German people. He expressed his genuine appreciation of all the courtesies and good will extended his representation, both civil and military, by the agents, both civil and military, of the States represented in the High Commission. He asserted that it had been a great privilege and pleasure to be associated in the labors of the High Commission and that in coming years he would take pleasure in thinking of the historic period in which America participated in the watch on the Rhine.

The American delegation withdrew at the conclusion of these addresses. This was the last sitting of the High Commission attended by any American representative.

- - - 000 - - -

---

[1]See Appendix 3: Addresses by Monsieur Tirard, Baron Rolin-Jacquemyns, Lord Kilmarnock and General Allen.

1697[d]
Interallied Rhineland High Commission.

Office of American Representative.

Coblenz, April 11, 1922.

In re: Appeal of Joseph Smeets of March 25, 1922.

Herr Smeets appeals to the High Commission (1) to remove his case pending before the Amtsgericht, Cologne, from German jurisdiction; and (2) to delay execution of the sentence until the High Commission acts thereon.

He invokes (1) Ordinance No. 70, Paragraphs 3 and 4;
(2) Ordinance No. 1, Article 15 (2), Caluse 2; and
(3) Ordinance No. 2, Article 17.
as a basis for his appeal.

Approaching this issue firstly with reference to the relevancy of citations (2) and (3) supra. Since the appeal of Herr Smeets is from "a conviction and judgment" of the Amtsgericht, Cologne, of February 13, 1922, sustained on appeal by the Landgericht (Criminal Court No. 6) on March 10, 1922, it is obvious that the proceedings against Herr Smeets were of a criminal character. If this be so, the appeal under Ordinance No. 2, Article 15, Paragraph 2, Clause 2, and under Ordinance No. 2, Article 17, is not receivable, because the Articles in question have reference to civil causes, civil jurisdiction, and are not intended to apply to proceedings in a German criminal court.

This view is potentially emphasized if recourse be had to Article 18 of the same ordinance, referring to the institution and jurisdiction of the "High Commission Court". Section 3 of that Article reads:

> "It shall be the duty of these Courts to decide any civil cases referred to them under the provisions of the proviso to Clause 2 of Article 15 above ......"

This "proviso to Clause 2 of Article 15 above" is one of the citations relied upon by Herr Smeets to support his applications to the High Commission, but it manifestly avails him nothing, since his appeal is from a criminal prosecution and judgment, while the jurisdiction of the High Commission Courts is specifically civil. It may be added, as a matter of sound legal construction, that the additional wording of the proviso to Clause 2 of Article 15, to wit:

> "or disposed of in such way as the High Commission may direct" ....

was not intended to change the character of issues or actions covered by Article 15 of Ordinance No. 2, for this Article is the first enumerated under Part II of that ordinance dealing with "Civil Jurisdiction".

The words above indented and quoted show merely a reservation of power in the creating body itself, the High Commission, to deal with the "Classes of cases" referred to in proviso to Clause 2, otherwise than by reference to its High Commission Courts; but it never was intended to widen the jurisdiction or change the character of the actions embraced in Article 15, which exclusively pertains to civil jurisdiction and civil actions.

Moreover, this proviso to Clause 2 of Article 15 of Ordinance No. 2 confers original and not appellate jurisdiction. The prohibition that certain cases "shall . . . not be heard before the German Courts but, etc.", clearly indicates the intent of the High Commission. Under such a construction, the appeal of Herr Smeets is again not receivable by the High Commission.

In conclusion on this point, it may be observed that the uniform practice of the High Commission in construing Articles 15 and 17 of Ordinance No. 2 has been to confine their operation or applicability to protected persons, enumerated in Article I and Clause 5 of Article II of Ordinance No. 2, i.e., to the High Commission and Army personnel. (See Mohu case, Legal Committee, File 344). Herr Smeets is not a person embraced within these classifications. Accordingly no legal reason appears to justify Herr Smeets' appeal based upon Articles 15 and 17 of Ordinance No. 2.

Reverting to Herr Smeets' appeal for intervention by the High Commission under Ordinance No. 70, and assuming, for the sake of argument, that his conviction was for a political offense, paragraph 3 of that ordinance is not applicable. The date of commission of his alleged offense is given by him "November and Décember 1921"; but the paragraph cited refers explicitly to a "political act done during the period of the Armistice." That period ended with the entry into vigor of the Treaty of Peace, January 10, 1920. Hence, no "political" act done or committed subsequent to that date would entitle him to immunity under paragraph 3, Article I of Ordinance No. 70, or warrant his invocation of the High Commission's protection.

Herr Smeets, while denying his alleged guilt, remarks that the acts charged

> "would denote a political offense on my part in close connection with the policy inaugurated by me during the Armistice ...... In this case therefore the necessary conditions for the intervention by the I.A.R.H.C. in accordance with the ordinances quoted above would be fulfilled."

If by his reference to a "policy inaugurated by him during the Armistice", Smeets hopes to establish a causal connection between that policy and the act for which he has been convicted, so as to make the "policy" and the "act" one and, accordingly, an Armistice transaction, his reasoning is legally unsound. It is a firmly established principle of law that "intent" unaccompanied by an overt criminal act is not punishable; but the moment the overt act is committed, criminal process may be invoked; and, for all statutory purposes, the time of the commission of the act controls, irrespective of how long previously the intent may have existed, for it is merged in the act itself.

An illustration of this principle is seen in a conspiracy, where two or more persons combine to commit some offense or punishable action. Although the statements made by these inter sese, or the plans outlined, discussed or approved by them jointly, may become known to the authorities, no prosecution against them or either of them may be based on such evidence, nor can the same be used against the conspirators or either of them until proof of the commission of some overt act toward the accomplishment of their common design is first established. The "policy" of Herr Smeets may truly enough have existed, in either an active or dormant sense since the period of the Armistice, and yet not render him liable to prosecution. On the other hand, as soon as he had done something, which the German authorities considered criminal, proceedings against him would be in order and the only time then to be considered would be the date of such action by him.

It seems that there is an analogy between the "policy" of Herr Smeets and his associates and the illustration quoted, to-wit: a conspiracy. Political movements for the overthrow of an existing governmental system partake of the nature of a conspiracy and their projectors are referred to as "political conspirators". Their views, policies, etc., not translated into physical action, are not punishable, but the moment they lead to such, prosecution follows. But the acts of recent occurrence for which Smeets is prosecuted are not even parts of a connected whole as in cases of conspiracy. The most that can be said is that Smeets' motive has been continuous. The acts for which he is prosecuted are entirely separate acts.

It is accordingly irrelevant, immaterial and inconsequential as to when the "views" or "policies" were first adopted. The criminal act is the gravamen to consider and punish. In accordance with this view, Herr Smeets cannot relate back an act committed in "November - December 1921" to a date prior to January 10, 1920. His invocation of Ordinance No. 70 is therefore not receivable by the High Commission.

Granting the validity of the arguments above, it may nevertheless be contended that Herr Smeets should be protected because he is being prosecuted "on account of services rendered" to the "Allied authorities of occupation" or on account of "relations" with said authorities within the meaning of Ordinance 90. The relations and the services contemplated by Ordinance 90 are those with "the Allied authorities of Occupation". Not all relations with Allied persons in the Occupied Territories constitute relations with "Allied authorities of Occupation". Clearly, the ordinance refers to services and relations which assist the Allied authorities of occupation in the performance of their duties under the Rhineland Agreement. Herr Smeets contends in his appeal that his alleged offense was a political offense connected with a policy inaugurated by him during the Armistice. Have the Allied authorities in performing their duties under the Rhineland Agreement any relations with such an offender, or is the commission of such an offense a service rendered to the Allied authorities of occupation in the performance of their duties? Herr Smeets' political policy is well known. It contemplates at most the separation of the Rhineland Territory from Germany, and at least the separation of the Rhineland Territory from Prussia. Would the performance of such a program perform or violate the Rhineland Agreement? In Article 5 of this Agreement we read that the civil administration of the Occupied Territories shall remain in the hands of the German Authorities and the civil administration of the areas shall continue under German law and under the authority of the Central German Government, except insofar as it may be necessary for the High Commission to adapt that administration to the needs and circumstances of military occupation. When the armies of the Allied authorities came into these territories, the German military forces withdrew. It might be contended that the Allied armies thus succeeding, succeeded also to the duty of the German armies to see to it that lawful governments be not overthrown. It might be contended that Article 5 makes it the obligatory duty of the Allied authorities to see to it that the authority of the Central German Government shall continue and that the local German Government be not overthrown. Certainly the Article should prevent the Allied authorities from protecting persons who conspire to overthrow these governments when such persons are prosecuted for their unlawful acts.

(Signed) HENRY T. ALLEN.

THE SMEETS CASE

III.

SUSPENSION OF SENTENCE PASSED AGAINST HERR SMEETS

Annex 2060. (File 7570/1).

(Reference Minutes 3731 of the 150th Sitting).

2838 The President asked the Deputy British High Commissioner whether he had received instructions permitting him to examine the Smeets case, discussion of which had been postponed at the request of his High Commissioner.

The Deputy British High Commissioner replied that he was prepared to discuss the case and desired that it be decided if possible at the present Sitting, for, leaving aside the personality of Smeets, the principles behind this question were all important.

In order to avoid any possibility of misunderstanding the views of the French High Commissioner he would wish to put the following three questions:-

1. The offences in respect of which Smeets had been sentenced were a number of cases of criminal libel and one case of causing newspapers to be sold in public contrary to German law. These offences were committed during 1921 and 1922. According to British legal conception and indeed, as far as he knew, according to the legal conception of most countries, the date of an offence was the date on which the act constituting the offence was committed and not the date on which the intention to commit the act was formed. As he understood the French argument, since Smeets commenced his separatist policy during the period of the Armistice the abovementioned acts committed in pursuance of this policy were held to come under the provisions of Ordinance No. 27 as modified by Ordinance No. 70 and therefore not punishable. He would like to know where the line was to be drawn in respect of acts alleged to have been committed in pursuance of this policy. If to further his political ends Smeets robbed a Government office or maltreated a Prussian policeman, would he be held to be immune?

2. Was the proposal that Smeets should be protected dictated by legal scruples based on the Ordinances of the High Commission or by political considerations?

3. If the proposal to protect Smeets were to be adopted, what action was it suggested that the High Commission should take? Were the sentences passed on Smeets to be completely suspended? The High Commission would doubtless realize that if Smeets were granted complete immunity in respect of these actions not only he but all his party would have good grounds for believing that they would be immune in the future in respect of any offences which they asserted to have been committed to further their political ends.

The French High Commissioner, replying to questions 1 and 3 stated that in his view there could be no question of taking a decision which should cover all the future acts of Smeets or bind the High Commission in respect of other persons. That was a question of fact to be examined in connection with the attitude of the German authorities in each specific case.

With regard to the second question, the case should be examined both from a legal and from a political point of view. The spirit of the decisions of the Allied Governments and of the Ordinances of the High Commission in matters of this kind was in fact as much political as legal. The Allied Governments, even before they set up the High Commission, undertook to protect after the entry into force of the Peace Treaty, persons who might be subject to molestation by the German authorities as a result of their political activities in the Occupied Territories during the Armistice period.

He would point out also that whilst the German authorities showed great tolerance towards certain German newspapers which published very violent attacks against the Government they had during recent months taken proceedings against and convicted Herr Smeets uninterruptedly on manifestly unimportant charges.

Finally, it was a striking fact that official authorities in Bavaria had assumed towards the Reich an attitude of rebellion which had led neither to prosecution nor to punishment, whereas Herr Smeets was persecuted for acts which were trifling in character and out of all proportion to the Bavarian activities.

It was therefore clearly proved that the German authorities were not inspired by a scrupulous desire to do justice but rather by political considerations the object of which was to paralyze Herr Smeets' activities, which tended at the moment towards a rapprochement with the authorities of occupation, by depriving him of his liberty. It was due to Herr Smeets' having declared publicly - perhaps incorrectly - that he was the friend of the French, British and Belgian authorities that he had been made the subject of repeated proceedings. This case therefore definitely fell within the scope of Ordinances Nos. 70 and 90 and at the stage which the question had now reached it appeared impossible that the High Commission could, without showing weakness, fail to repress a manifest attack upon its prestige.

With regard to the third question, concerning more especially the point whether the sentences should be suspended completely or only in part, the French High Commissioner, before replying, asked his British colleague whether he felt that discussion of this point might lead to unanimity among the High Commissioners. If so, he would gladly bring it forward.

The Belgian High Commissioner declared himself entirely in agreement with the opinion expressed by the French High Commissioner. He added that, in his opinion, it was due to the attitude of the German courts that this question had assumed a political complexion. In proceeding against Herr Smeets they had, it appeared, wished to attack the Allied Authorities; he considered that in these circumstnaces it was the duty of the High Commission to forbid the execution of the sentences passed.

The Deputy British High Commissioner replied that he could not agree that the High Commission had any grounds for intervention under the provisions of Ordinances No. 90. The sentences, which amounted to 8 months and two weeks imprisonment, did not appear to be excessive in view of the number and nature of the offences of which Smeets had been convicted.

As, according to his views, neither Ordinance No. 27 nor Ordinance No. 90 offered the High Commission any legal basis for intervention, the only decision that, in his opinion, the High Commission could take would be to allow the immediate execution of the sentences.

His view being this, and the views of the Belgian and French High Commissioners, as he understood them, being that no sentence of imprisonment should be served by Smeets, he feared there was little chance of unanimity being reached along the lines suggested by the French High Commissioner.

Colonel Stone stated that the American Representative felt that the High Commission could not legally interfere with the execution of the sentences pronounced against Josef Smeets by German courts; in his view the Ordinances upon which Herr Smeets relied in his appeals were not applicable and no Ordinance would justify the High Commission in according Smeets any protection in the cases under consideration.

The French High Commissioner referred to the text of Ordinance No. 90 and explained how in his opinion it was incontestably applicable to the case of Herr Smeets.

Nevertheless, as a measure of compromise he proposed to his colleagues that the question be postponed sine die and referred to the Allied Governments.

The Deputy British High Commissioner replied that on the one hand this procedure would be equivalent to a decision in Smeet's favor and that on the other hand his Government desired that the matter should be settled by the High Commission.

The French High Commissioner proposed as a further measure of compromise and with a view to establishing unanimity that the previous decision of the High Commission, which provisionally suspended the execution of the sentences, should be maintained.

The Deputy British High Commissioner stated that he had the same objections to this as to the previous proposal. He viewed with grave concern the decision put forward, which, he was convinced, was a mistaken one and whose effects would be far-reaching. He requested therefore that a full record of the discussion be made in the Minutes of the High Commission in order that it might be clearly seen upon whom the responsibility for the decision rested.

The President then put the following resolution to the vote:

Whereas Herr Josef Smeets has in due form laid before the High Commission successive appeals in connection with a series of proceedings brought against him before the German courts of justice;

As regards the facts of the case:

Whereas it results from the documents in the case and from the circumstances of the proceedings that Herr Smeets, charged for the most part with offences against the laws governing the press, has been prosecuted by the German courts of justice for a large number of acts, the majority of which are of a minor character and would not have led to prosecution unless account had been taken of the personality of their author and the political campaign conducted by him against the Prussian Government during the period of the Armistice:

And Whereas Herr Smeets has on several occasions and even recently, both in the press and in political meetings, publicly manifested the desire to entertain good relations with the Authorities of Occupation and to bring about a rapprochement with the occupying nations:

As regards the law applicable to the case:

Whereas the Allied Governments, in a Note dated July 27, 1919, undertook to protect after the entry into force of the Peace Treaty persons who might be subject to molestation as a result of their political activities in the Occupied Territories during the Armistice period and whereas the High Commission has given force of law to this protection in its Ordinance No. 27, amended by Ordinance No. 70:

And Whereas the intention of the High Commission was, in the interests of the safety of the Armies and the maintenance of public order, to protect persons subjected to molestation by reason of services rendered or alleged to have been rendered to the Armies of Occupation or by reason of their real or supposed relations with the said Authorities, and whereas the High Commission has given legal effect to this intention by its Ordinance No. 90;

<u>And Whereas</u> it results from the considerations set forth above that the High Commission has good reason to hold Herr Smeets entitled to claim the benefit of the provisions of the abovementioned Ordinance;

<u>And Whereas</u> the very number and the nature of the sentences passed clearly prove that in this matter the German authorities have not been inspired by a scrupulous desire to do justice but rather by political considerations with the object of punishing Herr Smeets for what they regarded as transgressions committed by him in the past and to paralyze, by depriving him of his liberty, his present activities which tend to a <u>rapprochement</u> with the Occupying Authorities:

THE HIGH COMMISSION DECIDES AS FOLLOWS:

No steps shall be taken to proceed with the execution of the sentences passed against Herr Smeets on February 15, 1922 (Case 1 J 690/21, March 3, 1922 (Cases 1 J 347/21 and 1 J 506/21). June 12, 1922 (Case 1 J 594/21 and June 19, 1922 (Case 1 J 772/21), and referred to in letters Nos. 5418/H.C.I.T.R.[1] dated February 12, 1922 and 5923/H.C.I.T.R.[2] dated April 13, 1922.

Monsieur Tirard (France) and Baron Rolin Jacquemyns (Belgium) voted for and Lt. Colonel Ryan (Great Britain) against this resolution.

The resolution was declared carried by a majority and would be notified for execution to the German Commissioner[3] after adoption of the Minutes of the Sitting.

---

[1] Annex 1462c of the 122nd sitting.

[2] Annex 1605d of the 130th sitting.

[3] Annex 2066b.

Annex No. 2006

File 7570/1.

Translation.

2066b Interallied Rhineland High Commission.

Coblenz, October 7, 1922.

No. 7549/H.C.I.T.R.

The President of the Interallied Rhineland High Commission

to

The German Commissioner for the Occupied Rhineland Territories, Coblenz.

In reply to your letter No. I.4423, dated August 24, 1922, regarding the sentences passed against Herr Smeets, of Cologne, I have the honor to communicate to you the following resolution adopted by the Interallied Rhineland High Commission:

Whereas Herr Josef Smeets has in due form laid before the High Commission successive appeals in connection with a series of proceedings brought against him before the German courts of justice:

As regards the facts of the case:

Whereas it results from the documents in the case and from the circumstances of the proceedings that Herr Smeets, charged for the most part with offences against the laws governing the press, has been prosecuted by the German courts of justice for a large number of acts, the majority of which are of a minor character and would not have led to prosecution unless account had been taken of the personality of their author and the political campaign conducted by him against the Prussian Government during the period of the Armistice:

And Whereas Herr Smeets has on several occasions and even recently, both in the press and in political meetings, publicly manifested the desire to entertain good relations with the Authorities of Occupation and to bring about a rapprochement with the occupying nations;

As regards the law applicable to the case:

Whereas the Allied Governments, in a Note dated July 27, 1919, undertook to protect after the entry into force of the Peace Treaty persons who might be subjected to molestations as a result of their political activities in the Occupied Territories during the Armistice period and whereas the High Commission has given force of law to this protection in its Ordinance No. 27, amended by Ordinance No. 70;

And Whereas the intention of the High Commission was, in the interests of the safety of the Armies and the maintenance of public order, to protect persons subjected to molestation by reason of services rendered or alleged to have been rendered to the Armies of Occupation or by reason of their real or supposed relations with the said Authorities, and whereas the High Commission has given legal effect to this intention by its Ordinances No. 90;

And Whereas it results from the considerations set forth above that the High Commission has good reason to hold Herr Smeets entitled to claim the benefit of the provisions of the above-mentioned Ordinances;

And Whereas the very number and the nature of the sentences passed clearly prove that in this matter the German authorities have not been inspired by a scrupulous desire to do justice but rather by political considerations with the object of punishing Herr Smeets for what they regarded as transgressions committed by him in the past and to paralyze, by depriving him of his liberty, his present activities which tend to a rapprochement with the Occupying Authorities:

THE HIGH COMMISSION DECIDES AS FOLLOWS:

No steps shall be taken to proceed with the execution of the sentence passed against Herr Smeets on February 15, 1922 (Case 1 J 690/21, March 3, 1922 (Cases 1 J, 347/21 and 1 J 506/21). June 12, 1922 (Case 1 J 594/21) and June 19, 1922 (Case 1 J 772/21), and referred to in letters Nos. 5418/H.C.I.T.R.[1] dated February 12, 1922 and 5923/H.C.I.T.R.[2] dated April 13, 1922.

You are requested to notify this decision, for execution, to the German judicial authorities concerned.

The President of the Interallied Rhineland High Commission:

(Signed) P. TIRARD.

The Interallied Secretaries General:
(Signed) A. ISAMBERT. - W. H. FOX.

---

[1]Annex 1462c of the 122nd sitting.

[2]Annex 1605d of the 130th sitting.

APPENDIX 3.

Appendix No. 3 Chapter 2.

ADDRESSES DELIVERED DURING SESSION OF INTERALLIED RHINELAND HIGH COMMISSION ON FEBRUARY 6, 1925, ON THE OCCASION OF THE DEPARTURE OF GENERAL ALLEN, REPRESENTATIVE OF THE UNITED STATES GOVERNMENT.

Monsieur Tirard said:

Gentlemen:

Before opening the Sitting I must call the attention of my colleagues to the fact that today for the last time our American friends are present at this table. General Allen has been good enough to honor us with his presence. I wish to thank him, both as President of this Assembly and as French High Commissioner.

It is unnecessary to dwell upon the military merits of General Allen. We know how magnificently the American garrison at Coblenz bore itself and that such troops can only be commanded by a great chief. I wish rather to express for my part my opinion of the diplomat and the man. In respect of the former we must cast our thoughts far back. You will remember doubtless, gentlemen, the difficult circumstances in which the High Commission assumed its functions at the end of 1919 and the beginning of 1920. From that date the United States Government expressed their desire to dissociate themselves somewhat from European affairs and did not ratify the Treaty of Versailles. Perhaps the event which brings us together now is but the normal consequence of a policy which it does not behoove us to judge. But from that moment General Allen found himself in a situation which was, from a diplomatic point of view particularly delicate. The Allied Governments having promulgated the Agreement, the High Commission had to exercise its authority over the whole of the Occupied Territories of the left bank of the Rhine. On the other hand the United States Government, not having ratified the Treaty of Versailles but having most happily maintained its troops at our side, the Coblenz area remained under the Armistice ~~regime.~~ A solution had to be found to this difficult problem. It is thanks to the courtesy and diplomacy of General Allen that the American Representatives took their place at our deliberations. It is thanks to him that for three years in the most varying circumstances no difficulty has arisen from the special situation of the American troops. In this zone our Ordinances, drawn up of common accord with the American Representatives, were promulgated by General Allen as military orders and for three years our relations have been of the best.

As for the man, I will express what you all feel; every time that he has taken his place amongst us we have been impressed by his high-mindedness and the perfect impartiality of his judgments, to which we have always endeavored to give the greatest possible weight. It is not without real emotion that I salute for the last time the Representative of the United States of America. We shall always associate the remembrance of General Allen with the magnificent spirit which, in the hour of danger, united the troops of his country and those of the Allies in their struggle for the identical principles of justice and liberty.

I address at the same time my thanks to the whole of the American delegation and particularly to Colonel Stone and to Major Davis who have both lent their valuable collaboration in the work of the High Commission and whose labors are incorporated in our Ordinances and Decisions.

Baron Rolin-Jacquemyns, Belgian High Commission, said:

I will content myself with asking your permission to say in a word that I associate myself most heartily in my name and in the name of my collaborators with what you have said about the merits of General Allen and the services which he has rendered.

Not only was America here at our side but, in expressing the desires and the sympathy of the American people, General Allen was better able than any other to express the feeling of union which has never ceased to reign amongst us.

I thank you for it and I will not forget.

Lord Kilmarnock, British High Commissioner, said:

I would like to add a word on my own behalf and on behalf of the British Department to say how much we have always appreciated the tact and courtesy which General Allen has consistently exhibited and the spirit of high-minded impartiality with which he has handled the many delicate questions which have from time to time arisen. It has been a great pleasure to have him amongst us and we have all appreciated his high qualities. It only remains for me to express my extreme regret that this is the last time that General Allen will be amongst us at the Sittings of the High Commission.

General Allen replied:

Mr. President and Gentlemen:

The brief announcement of the reason for the severance of our relations with the Rhineland and the High Commission was contained in my letter addressed to you, M. Tirard, as President of the High Commission under date of February 2nd. Now I am here at your request to associate myself in this session with the members of this Superior Body for the last time. I have no words with which to properly express my personal gratitude for the great courtesy, consideration and unfailing generosity you have at all times extended to me. In making this statement, I am also the interpreter of the feelings of all my collaborators who have had the honor of personal contact with you.

After a period of four years on the Rhineland in charge of the American zone and nearly three years at the High Commission, ties of friendship have been established that take precedence over any discordant thoughts that may have arisen during this long period. It is but natural that far away United States should at times view measures of transcendental importance to the countries directly concerned, through American glasses, but our close association here with you as principal actors in measures of world importance, especially during recent times, has been potent in its enlightening influence.

We have worked in common counsel towards the solution of many difficult and delicate problems coming before the High Commission in regard to the government of the Occupied Territories and the relation of our military forces to the German people Our deliberations have been conducted in two languages under the handicap of dissimilar laws, traditions and customs of our respective countries, and yet so great has been the spirit of tolerance, kindness and friendly cooperation which has governed each member of this body that our differences have been compounded and we have been able to reach an agreement on almost every question. Every law and decision enacted by this body has been made for the four zones alike and has found application throughout the Occupied Territories.

The many complicated questions connected with state administration and policy must serve to determine the course of the respective states. To criticize these would be an act of presumption foreign to my thoughts, but to express appreciation of all the courtesies and good will extended our representation both civil and military by the high agents both civil and military, of the states represented on the Commission, gives me genuine pleasure.

It has been a great privilege to share in your labors and to benefit by your wisdom and experience in working with you in common purpose towards the execution of our missions. In the delightful memories of coming years and in the reveries of this historic period in which America participated in the Watch on the Rhine, each of you will be in my mind in affectionate remembrance.

General Allen and the American delegation then withdrew.

The President declared the Sitting open.

ANNEX

ORDINANCES OF INTERALLIED

RHINELAND HIGH COMMISSION

Jan. 1, 1922 - Feb. 2, 1923.

(Photograph omitted)

Identifications
Photograph of Interallied Rhineland High Commission Personnel.

First Row (bottom) - Left to Right.

1. Commandant O. Vermeulen, Belgian Member, Requisitions, Coal and Communications Committees.
2. Mr. F. Vibier, Belgian Financial Adviser.
3. Comte Raoul de Liedekerke, Belgian Deputy High Commissioner.
4. Mr. Amedee Roussellier, French Deputy High Commissioner.
5. Commendatore Dell'Abbadessa, Italian Representative.
6. Baron E. Rolin-Jacquemyns, Belgian High Commissioner.
7. General Henry T. Allen, American Representative.
8. Mr. Paul Tirard, French High Commissioner, President IARHC.
9. Lord Kilmarnock, British High Commissioner.
10. Colonel David L. Stone, Representative of General Allen on High Commission.
11. Lt. Col. R.S. Ryan, British Deputy High Commissioner.
12. Mr. Manton Davis, American Legal Adviser.
13. Mr. H. Maxwell Thin, British Legal Adviser.
14. Mr. F.H. Quarry, Intelligence Officer, British Department.
15. Mr. A. Michelson, Assistant British Legal Adviser.

Second Row (top).

1. M. Poisson, Representative of French Treasury Department attached to High Commission.
2. M. Ehrmann, clerk to Secretary General (French).
3. Mr. R. de Bouille, French Department Interallied Secretariat.
4. Mr. A. Dorff, District Representative for the Belgian zone of Occupation.
5. Mr. J. de Codt, Representing Belgian Department, Economics and Communications Committees.
6. Mr. de Courson, Secretary Administrative Committee.
7. Mr. M. Silhol, Chief Assistant to French High Commissioner.
8. Mr. Roussel, French Legal Adviser.
9. Mr. G. Lambotte, Belgian Legal Adviser.
10. Mr. E.V. Gascard, Chief of Finance, French Department.
11. Commandant F. Carteron, Representing French Department on Economics Committee.
12. Mr. A. Isambert, Interallied Secretary General (French).
13. Mr. W.H. Fox, Interallied Secretary General (British).
14. Lt. Col. C. d'Arbonneau, Chief of French Military Bureau.
15. Mr. C.F. Madeley, Private Secretary to British High Commissioner.
16. Mr. E. Troughton, Financial Adviser and Administrative Officer, British Department.

17. Mr. J.J.M.Herbertson, Assistant Intelligence Officer, British Department.
18. Mr. J.M.Butters, Assistant to Administrative Officer, British Department.
19. Mr. E. Wareing, Assistant Secretary General (British).
20. Miss Jennings, Minutes Clerk, British Department.
21. Mrs. J.M.Butters, Clerk to Secretary General, (British).
22. Mr. A.P.Happel, Assistant Legal Adviser, American Dept.
23. Mr. H.F.Angold, Agenda Secretary (British).
24. Mr.H.L.Kingston, Representing American Department, Finance and Administrative Committees.

CHAPTER III

LIAISON BETWEEN AMERICAN FORCES AND POPULATION.

Sub-Chapters | Page

----- 000.0 000 -----

SUB-CHAPTER A.

OFFICE OF CIVIL AFFAIRS.

Re-organisation of Office.

No material changes were made during the year in the general policy relative to relations between the A.F.G. and the civilian population in the American area. The reduction in strength of the military forces, however, necessitated a corresponding reduction in the personnel of the Office of Civil Affairs and a re-organisation of the office. This resulted in the following division of duties:

Designation	Function
Officer in Charge of Civil Affairs	General supervision and policy- legal decisions- reviews of cases tried by superior Provost Court.
Executive Officer	Liaison with German officials and civilians - custody of all records - coordination.
Courts	Administration of Superior Provost Court - collection of funds derived from fines imposed.
Department of Sanitation and Public Health	Supervision of civilian health conditions and sanitation and conditions of jails in which prisoners sentenced by Superior Provost Court were confined.

The peculiar status of the American Forces in Germany necessitated the closest cooperation between the Office of Civil Affairs and the Interallied Rhineland High Commission, which was effectively maintained by having the Commanding General's unofficial representative on the Interallied Rhineland High Commission act also as the Officer in Charge of Civil Affairs.

Executive Officer.

The relations existing between the American Forces in Germany and the civilian population, as shown in the dealings between the Executive Officer of Civil Affairs and local officials, were most satisfactory. The German officials cooperated zealously with the Americans in maintaining order and enforcing such regulations as were issued affecting the German people. This spirit of cooperation and confidence was particularly evident from the latter part of 1922 to the date of the withdrawal of the troops. That it existed among all ranks and classes was shown by the decrease in the number of reports of conflicts between the members of the occupying forces and civilians - which could not be entirely attributed to the reduction in strength of the forces.

The American area was singularly free during the last thirteen months of our occupation, from disorders of all kinds. The only incident which threatened serious results and required the interference of the American authorities, outside of the Military Police, occurred at Bendorf on July 15th and 16th, 1922, when, after a dispute between the labor elements and the management of the Concordia Huette, upon which the city of Coblenz was dependent for its gas supply, a walk-out resulted, causing the complete cessation of all work in the plant. Ordinance 53, I.A.R. H. C., made effective in the American area in General Orders No.

91, Headquarters, A.F.G., November 2nd, 1920, required that before a strike or walk-out became effective, all labor disputes affecting utilities necessary to the maintenance, safety or requirements of the armies of occupation be submitted to the authorities prescribed by German law. Since this procedure had not been followed, the Commanding General designated an officer to proceed to Bendorf and inform the laborers that they must resume work and submit their dispute through proper channels. While a sufficient number of laborers complied with this order to resume operation of the blast furnaces, three labor leaders were arrested, and the workers struck in protest on the following day. Troops were immediately sent to Bendorf and sufficient labor was requisitioned to insure the operation of the furnaces, the entire personnel of the plant resuming work the next day.

The military authorities established no civil courts, provision being made for utilization of the German civil courts when appropriate. In order to prevent trial before these courts of the many trivial cases of disagreement between merchants, etc., and members of the occupying forces, the Executive Officer acted as mediary to bring about amicable settlements. It was a rare exception when an agreement acceptable to both parties was not reached.

When it became known among the civilian population that the American Forces in Germany were about to be withdrawn, the number of complaints and requests increased greatly. These might be classified generally under the following headings:

> Claims for support of illegitimate children whose fathers were alleged to be members of these forces;

b. Unpaid bills against members of A.F.G.;

c. Controversies between members of A.F.G. and merchants, doctors, etc., submitted to Executive Officer for adjustment, and

d. Requests for work and charity.

In addition to these, and considerably in the majority, were the pleas for clemency for prisoners sentenced by the Superior Provost Court which, in exceptional cases, were acted upon by the Executive Officer, in most cases, however, submitted to the Pardon Board for action.

<u>Courts.</u>

The Superior Provost Court tried many more cases of theft in 1922-1923 than in the preceding years. These varied from minor offenses, such as stealing small quantities of foodstuffs, to burglary and robbery on a large scale. The basic cause of the act in many instances was the same, viz., actual want, which resulted from the depreciation of the German mark and the accompanying increase in the cost of living, with wages keeping in the background.

Due to the same reason, an increase was noted in the number of cases tried that involved the sale of stupefying drugs to members of the occupying forces. Drugs could be obtained cheaply in Germany and trafficking therein offered large profits. The American soldier, due to the rate of exchange, was bountifully supplied with money, and was therefore the favorite prey of drug-brokers, who portrayed the profits obtainable should the drugs be sold by the soldier in France or Belgium.

There was, however, a decrease in the number of vagrant women tried, a fact largely accounted for by the reduction of effectives of the American forces. The policy of deporting vagrant women arrested in the American area, but whose homes were outside of the area, prevented too great an influx of undesirable characters. Violation of deportation orders was an offense punishable by imprisonment. This policy with reference to vagrant women received the whole-hearted support of the German officials, welfare societies and civilians.

From January 1st, 1922, to January 15th, 1923, a total of 2,345 cases was tried by the Superior Provost Court, 299 thereof being acquittals and 2,046 convictions. The convictions were as follows:

Sale of prohibited alcoholic drinks	98
Unlawful possession of Government property	62
Failure to obey lawful orders	55
Theft and similar offenses	163
Obstruction	4
Assault	31
Insult	10
Unlawful use of uniform or badge	12
Unlawful possession of fire arms	85
Offenses against circulation regulations	81
Discrimination against military persons.	9
Holding unauthorized meetings	1
Traveling in reserved military compartments	11
Unauthorized display of flag	1
Failure to mark prices legibly	29
Aiding or abetting crime	13
Trespass	24
Violation of deportation orders	3
Overcharging	14
Violation of traffic regulations	21
Unnatural sexual intercouse	2
Vagrancy and prostitution	1,317

Disposition of Prisoners.

Prior to the date of transfer of command, the cases of all prisoners serving sentences awarded by the American Provost Court were carefully reviewed by a board consisting of the Executive Officer, Civil Affairs, the Provost Court Officer and the Coblenz Kreis representative of the American Section, Interallied Rhineland High Commission, who was also a member of the Board of Pardons, to which formerly all pleas for clemency were submitted. As a result of the Board's recommendations, fifteen prisoners serving jail sentences on Superior Court charges were released from confinement and the unexecuted portion of their sentences was remitted. The money fine imposed in addition to a jail sentence was remitted in the cases of three prisoners and their term of confinement reduced by six months. In addition to these cases, sixteen women serving jail sentences on vagrancy charges were released and the unexecuted portions of their sentences remitted.

The following shows the prisoners left in confinement to serve out their entire sentence, all of whom were convicted of committing serious offenses, or whose conduct while in jail did not warrant granting clemency:

Karmeliterstrasse Jail	48 prisoners (male)
Diez Penitentiary	1 prisoner "
Neuwied Jail.	3 prisoners "
Fischelstrasse Jail	2 prisoners (women)

All prisoners of American nationality, except one, and all women serving jail sentences, except the two noted above, were released. The American retained in confinement was a negro who had been in Germany for a number of years but who still claimed American nationality. He was convicted of assaulting an American soldier, the assault almost resulting in the death of the soldier, and was sentenced to one year's confinement and a fine of 10,000 marks. The assault was so brutal and dastardly that clemency in his case was not warranted.

All persons detained in the Montabaur Reformatory for Young Boys, the Good Shepherd Convent, a reformatory for young girls, and the Frauenanstalt, Bendorf, a venereal hospital for women, were left to complete their terms of detention or their medical treatment. These numbered:

Montabaur Reformatory	5
Good Shepherd Convent	8
Frauenanstalt	89

In the cases of the Montabaur Reformatory and the Good Shepherd Convent, which were model institutions of their kind, this action was taken on the urgent recommendation of the German Welfare Society of Coblenz. The French authorities contemplated continuing the Frauenanstalt on the same basis as that on which it was maintained by the American authorities and desired that the women undergoing treatment complete the cure, as otherwise the area would be flooded with diseased women who would constitute a menace to the French troops.

Lists of the prisoners held pursuant to American Court orders and the charges and the dates on which the sentences would expire were sent to each of the jails and reformatories, with instructions that the prisoners be released on the proper dates without further orders from the American authorities. Copies of the lists and instructions were sent to the French authorities, the _Kreis_ representatives of the Inter-Allied Rhineland High Commission, who were the supervisors of jails, the German Police Commissioner of Coblenz and the German Welfare Society of Coblenz. As all prisoners sentenced by the American Provost Courts during the year had served their sentences under German jurisdiction, the faithful carrying out of the instructions was confidently expected.

Status of Frauenanstalt.

As noted above, the French authorities intimated their intention of continuing the Frauenanstalt on its past basis. However, as a definite statement to that effect could not be obtained from them, and as the American requisition for the hospital lapsed on February 7th, 1923, instructions were issued to the hospital authorities that, should the hospital be retained by the French, the unexecuted portions of the sentences imposed on the patients were to be remitted, effective on the dates they were pronounced cured by the hospital authorities. Should the hospital not be requisitioned by the French, the patients were to be turned over to the German authorities, effective February 7th, 1923, and might be retained in the hospital until cured if the Germans decided to continue operations of the hospital. Should it be abandoned, the unexecuted portions of the sentences of the patients were to be rescinded and the patients released. Under no condition would any patient be required to serve any part of her sentence in jail.

Department of Sanitation and Public Health.

The activities of this department will be covered in full in Sub-Chapter C.

Close of Activities.

The activities of the Office of Civil Affairs, which had been continued in their entirety throughout 1922, officially ceased at noon on January 27th, 1923, at which time the Commanding General, American Forces in Germany, relinquished command of the American occupied area. Up to that hour, the Executive Office of Civil Affairs had dealt with the various requests and complaints of the civilian population as a liaison office with the military authorities, the Superior Provost Court had been in daily operation and the Department of Sanitation and Public Health had maintained surveillance over public health and sanitory conditions in the area.

Officers.

The following roster of officers on duty with the Office of Civil Affairs shows the assignments to duty, arranged chronologically, for the period January 1st, 1922 - January 1st, 1923.

Name	Rank	Duty
Stone, David L.	Colonel, G.S.	Officer in Charge Civil Affairs.
Goodrich, P. H.	Lieut.Col., Inf.	Superior Provost Court.
Downs, S. D., Jr.	Major, F.A.,D.O.L.	Executive Officer.
Crea, H.B.	Major, Inf.	Superior Provost Court and Executive Officer.
Flynn, T.J.,	Major, M.D.	Chief Sanitary Officer.
Andrews, P.M.,	Major, A.S.,D.O.L.	Executive Officer.
Thiele, C.H.	Major, C.A.C.,D.O.L.	Superior Provost Court
Reasoner, M.A.	Major, M.D.	Chief Sanitary Officer

CONTRAVENTIONS OF THE ORDINANCES from January 1, 1922 to January 15, 1923

Offense	Con-vict-ions	Total cases	Fines No.	Marks	1	2	3	4	5	6 mos.	Other sentences	Average days imprison-ment.
Failure to obey lawful order	56	68	23	58850	21	1	6	3	1	0	29 m. - 20 d.	41
Theft and similar offenses	163	202	63	284400	24	17	23	9	1	29	315 m.	97
Obstruction	4	4	1	1250	0	1	0	0	0	0	1-10 d.	10
Assault	31	42	8	28000	2	6	4	7	9	8	2 yrs.	90
Unlawful possession government property	62	83	27	381450	14	4	6	2	1	8	75 m. 25 d.	65
Sale of spirits, etc. to troops	98	120	63	294300	15	4	9	0	0	8	100 m. 5 d.	5
Discrimination	9	9	9	57600								
Unauthorized use uniform and badge	12	18	4	11000	1	2	1	1	0	1	2 yrs.	70
Offenses against Circulation regulations	81	85	3	2550	0	0	4	1	0	0	800 d.	16
Holding unauthorised meetings	1	1	1	50000	0	0	0	0	0	0	-	-
Unlawful possession of arms	85	97	32	00750	14	15	19	6	0	6	-	54
Riding in reserved military compartments	11	16	11	5765	0	0	0	0	0	0	-	-
Transporting cargo without manifest	0	1	-	-	0	0	0	0	0	0	-	-
Unauthorized display of flags	1	1	1	500	0	0	0	0	0	0	-	-
Failure to mark prices legibly	29	35	29	430000	0	0	0	0	0	0	-	-
Insult	10	14	5	38400	2	0	0	0	0	0	17 m. 10 d.	53
Aiding and alotting committing of a crime	13	18	7	33750	1	2	4	2	0	3	1-10 d.	9
Trespassing	24	30	7	6900	6	2	0	3	0	1	45 d.	30
Vagrancy and prostitution	1317	1453	All convictions carried sentences of not less than two months and not more than six. If found not diseased, sentences were suspended during good behavior, excepting in cases of repeaters. Of the convictions, 472 were admitted to Bendorf Hospital with venereal disease.									
Violation of deportation order	3	5	-	-	0	2	0	3	0	0	-	84
Presenting false claim	0	2	-	-	0	0	0	0	0	0	-	-
Overcharging	14	20	13	52660	0	0	0	0	0	0	-	-
Violation of traffic regulations	21	21	21	67500	0	0	0	0	0	0	-	-
Unnatural sexual intercourse	2	2	1	10000	0	0	0	1	0	0	1 yr.	240

Appendix No. 2(A)

POPULATION OF AMERICAN OCCUPIED

AREA BY KREIS IN 1922

Kreis Coblenz-Stadt	60,064
Kreis Coblenz-Land	67,348
Kreis Cochem	41,160
Kreis Mayen	80,337
Kreis Adenau	24,871
Kreis Ahrweiler	49,954
Kreis Neuwied	97,368
Kreis Montabaur[1]	7,000
	428,093

[1]Only a part of Kreis Montabaur was in the American area.

SUB-CHAPTER B.

RELATIONS WITH CIVIL POPULATION

General Opinion of A.F.G.

Soon after the local German papers announced that the American troops were to be withdrawn from the Rhine, the remark was made by a prominent businessman of Coblenz, and, having been heard, corroborated by many of his associates, that the Americans had had less discord with the citizens of Coblenz and had been more gentlemanly in their relations with the population than even the peace-time German garrison. This was probably true, for the soldiers of the A.F.G., especially during the last thirteen months, bore themselves generally in a very correct manner. Drunkenness on the part of Americans had become rare, and, consequently, squabbles with the population were not the every-day affairs they had once been. Another reason for the remark, which the speaker probably thought of as almost extravagant praise, was that the Germans, knowing the Americans would be replaced by French troops, were in much better position to appreciate the considerate treatment which had been theirs ever since the beginning of the American occupation.

The officers and men of the A.F.G. had, of course, grown accustomed to Germans and their ways. Just as they thought of the Rhine as a part of the commonplace scenery which they accepted as fact and which they greeted with acclaim when returning from leaves to other parts of Europe, simply because it had become a part of their life, so had they become accustomed to the excruciatingly slow business methods of the local population, including the ever-necessary thirty to forty-five minutes' wait at the banks whenever they wished to change a few dollars into marks. Naturally enough they were interested in local politics, especially as it had always been suspected that the French were behind one of the parties most in disrepute in the Coblenz area. This being the case, the American was in a position to understand the German viewpoint better than he had been three years before, and was also able to discourse on subjects of common interest with the inhabitants with whom he came in contact - his billet owner, the business and professional man whom he had dealings with, etc.

The merchants of Coblenz were able to make frequent turnovers of stock as long as the Stars and Stripes were floating over Ehrenbreitstein. American officers and soldiers were wealthy, according to the German viewpoint based on the prevailing rate of exchange, and were not at all reticent about buying anything which pleased their eyes. There were literally no luxuries for the Americans in the Rhineland; practically everything they might desire was within reach of their pocketbooks, provided, of course, the exchange was in their favor. When the mark dropped in relation to the dollar, the American bought goods in quantity, and when the mark stablised or improved - and at such a time prices were always increased - he quit buying. As a result, the German merchant viewed with alarm any undue increase in the value of the mark. In other words, the life and trade of the Coblenz merchants lay practically entirely in the hands of the Americans.

The Stadt Theater of Coblenz, one of the oldest in Germany, was facing a deficit of about 2,500,000 marks in February, 1922. The management was not in position to maintain the theatre for the 1922-1923 season without outside support, since the city of Coblenz, which owned it, was financially unable to grant a subsidy. This situation was made known, and a collection authorized to cover the deficit and to provide for the purchase of a new curtain. Almost the entire shortage and practically all the money necessary to buy the curtain were provided in a short time by the American colony. In addition to this, the American forces, in accordance with custom, voluntarily provided Christmas donations in 1922 for the purchase of presents for hundreds of poor children of Coblenz and its suburbs. The Catholic nuns, in their quest for alms for the support of the secular orphanages of the city, always visited the American offices and homes first, as it was from them that 80% of their income was received. Everything taken into consideration, the Americans left a wonderful impression with the inhabitants not only of the American occupied area but also of all the rest of Germany, for the news of their generosity to the poor and needy was heralded far and wide.

Whenever an American subject transgressed against the rights of a German national, he was adequately punished in case he was found guilty. The burghers of Coblenz learned to regard the American sense of justice as a standard which might well be striven for by all European nations - and they knew at the same time that, should they be found transgressing against the American regulations, they would be just as adequately punished.

As before, each train leaving Coblenz with transports of soldiers bound for the base ports brought hundreds of spectators to the station. German women and girls, and sometimes even men, wept unaffectedly as their friends or sweethearts were leaving the Rhineland. The American in Coblenz had made friends, and not acquaintances - and the rank and file of the local population was really sorry to see him go.

Marriages.

The same conditions in regard to marriage prevailed through 1922 and the first month of 1923 as described in Chapter 4, A.R. in O.G., 1920-21. The same method of dealing with applications at Headquarters was continued, namely: no applications to marry were approved when the character of the women was shown to be questionable, or when it was shown that she would not be desirable as the wife of an American citizen; each enlisted man desiring to marry had to deposit $100.00 to cover the expenses of his wife to his new station in the States; no married men below grade 3, except under exceptional circumstances, were allowed to remain in the A.F.G.; and no more applications from enlisted men below grade 3 were approved each month than there was accomodation for on the transport sailing the following month.

Upon the decision of the War Department, in the spring of 1922, to retain only a small garrison on the Rhine, a great number of applications was received from men returning to the States for permission to marry. In order to be able to gauge the accommodations necessary to return the men with their families, Headquarters, A.F.G., issued a letter, under date of March

6th, requiring all men whose organizations were under orders to return to the States to submit their applications before March 25th, 1922. Again in April, 1922, when the War Department had decided to withdraw the whole forces, the question of accommodation came up and it was decided that applications must be submitted before May 1st. After the decision of the War Department to retain a small garrison in Germany indefinitely, applications were few, and these were carefully scrutinised before approval.

The following chart shows the number of applications received and the number approved during the whole period of this report. It may be seen that during the months from June to December, only thirty-six applications were received, of which eleven were approved:

Month	No. Received	Approved	Returned to U.S. Women	Returned to U.S. Children	Total
January	37	28	29	13	42
February	68	47	24	10	34
March	145	97	83	23	106
April	105	80	143	59	202
May	204	183	212	88	300
June	13	1	64	40	104
July	2	0	0	0	0
August	0	0	0	0	0
September	1	1	0	0	0
October	3	2	0	0	0
November	12	6	0	0	0
December	5	1	0	0	0
TOTAL	595	446	555	233	788

On January 11th, 1923, upon receipt of advice from the War Department that the A.F.G. was to be withdrawn immediately, the question of marriage of enlisted men again became an issue. As a result of the great number of applications received, Headquarters issued the following circular, and in order to be able to extend the information to all concerned with the least possible delay, had it inserted in the Amaroc News and some of the local German newspapers:

"CIRCULAR NO. 5 13 January, 1923

REGULATIONS GOVERNING MARRIAGE OF SOLDIERS.

1. The attention of all members of these forces is invited to the fact that the marriage of a foreign woman to an American does not in itself make the foreigner a citizen of the United States; Therefore, before being able to enter the United States, they must comply with all passport and other regulations concerning the entry of foreigners. Under the laws of the United States whenever the immigration quota alloted to Germany is filled it will be impossible for foreigners of German birth to enter the country. Except in the case of officers, warrant officers and enlisted men above Grade 4, no member of these forces is entitled to transportation of his dependents to the United States.

2. Any soldier of these forces who nevertheless marries, must assume the responsibility of transportation of his wife to the United States by complying with all provisions of the immigration laws and by supplying funds for her travel.

3. Any marriage application submitted will contain a statement showing that the provisions of this circular will be complied with.

By Command of Major General ALLEN

OFFICIAL W. W. HARTS,
S.G. TALBOTT
Adjutant General Chief of Staff"

Seventy-one applications for permission to marry were submitted to Headquarters between January 1st and January 22nd, 1923. Of these, 24 were disapproved.

Since many of the married men were unable to pay the transportation of their wives to the States, several of the latter were left in Germany until such time as their husbands might be able to send them travel money. American and German authorities cooperated most fully, however, to enable the wives to travel on the Army transports with their husbands whenever financial difficulties did not intervene. . . .

Illegitimate Children.

The economic conditions existing in Germany apparently caused a lowering of morals among certain classes of German women. Innumerable cases in which women had associated with soldiers and the results had been pregnancy arose during the year. These cases, as a rule, did not come to light until after the soldier had returned to the United States and had failed to contribute to the support of the woman and child. The only action possible in such cases was to write to the soldier's commanding officer requesting such voluntary action as the soldier desired to take.

This subject will probably be the source of much correspondence for some time to come. It may be appropriate to note that in many deserving cases, i.e., when the women were not practicing prostitution, the soldier acknowledged his paternity and evidenced a desire to contribute towards the support of the woman and child.

Offences by Soldiers.

The following offences against Germans were committed by members of the American Forces in Germany in the period January 1st, 1922, to January 27th, 1923:

Assault	42
Creating disturbance. . . .	5

Death by automobile accident	3
Injury by automobile accident	6
Smashing windows and damaging property. . . .	10
Unauthorized hunting. . . .	2
Miscellaneous complaints. . .	4

By far the greater part of the offences listed were committed prior to June, 1922, before the forces were reduced to the 1,200 officers and men comprising the final remnant of the A.F.G.

THE WITHDRAWAL OF THE AMERICANS

(Translation)

(From "Coblenzer Volkszeitung"-Centrum-Coblenz, Jan. 22, 1923)

In a few days' time the American Forces of Occupation will evacuate the Coblenz bridgehead and embark at Antwerp on the return voyage homeward. According to latest orders, the troops are to leave Coblenz on January 25th and embark at Antwerp on the 26th.

If this withdrawal were taking place under ordinary circumstances, and if it signified the normal expiration of the period of occupation in accordance with the peace treaty, it would, apart from the personal considerations mentioned below, be greeted with a sigh of relief as a liberation from the heavy burden which has been weighing upon the Rhenish population for the past few years. As it is, however, that population is looking upon the departure of the Americans with mixed feelings, realizing with dismay that it does not mean the end of foreign dominion, but only the close of a phase thereof and the transition to another one whose features are completely obscure and of which all that may be said is that it will not be better or less onerous than was the first. On the contrary, all indications point to the transfer of the military power in the Coblenz bridgehead to that one of our former enemies who, as recent events prove, has retained the mental attitude of war to the most acute and lasting degree and who, in order to enforce his claims, although they be morally and legally unfeasible, does not respect even the limits which the Versailles Treaty drew for the victors. This is the political aspect which the withdrawal of the Americans presents to the German point of view.

When the troops of the Union entered Coblenz on that dismal 7th of December, 1918, they appeared flushed with victory - indeed, as conquerors whose entry into the war had decided the contest of arms against us. We had already finished with the others, including those who are trying to pluck the fruits of victory on the Continent today; but, at the last moment, the overpowering and untapped energy of America wrested the final victory from our hands, enfeebled by four years of fighting. The first few months of the period of occupation were characterized by anti-German sentiments, fed by a systematic press compaign that had lasted for years, and relations with the population were at first governed by the frame of mind of troops coming from the battlefields. Those sentiments then changed, and the foreign guests soon realised that the character of the German people, and especially the Rhinelanders, was entirely different from the gross caricature which Northcliffe and his satellites had made of it. With this realization, the amiable side of the American nature rose to the surface, the German blood flowing through the veins of many of them having its effect, too. As time went on there thus came about a state of affairs which, considering the circumstances, may be considered tolerable and even friendly, although the dark aspects of the situation, such as usually accompany the occupation of a district by a large body of troops and which made themselves felt with particular force in this instance, should not be forgotten. One contributory factor therein was the important fact that the well-paid troops spent their money for the necessities of life or the satisfaction of their personal inclinations on the same generous scale as that on which they received it. As the business world profited considerably thereby, the occupation was an advantage from the point of view of private enterprise, although a disadvantage as far as the coffers of the state were concerned, for the simple reason that Germany had to foot all the bills. Nevertheless, the money remained in the country and was even used for the relief of poverty and distress, children, particularly, profiting from liberal donations. The absence of all these factors will be felt considerably now that the Americans are leaving.

General Allen, Commanding General of the American Forces, and his officers, have done no mean service in establishing tolerable relations between the forces of occupation and the civilian population. Their correct attitude has showed their endeavor to grant the utmost possible freedom to political and private activities and, as far as consistent with instructions, the necessary obligations toward the Allies and the purposed of the occupation, to refrain from meddling and from pernicious interference in economic affairs. When General Allen goes he will be entitled to the same tribute as was paid by Presidents Thiers and MacMahon of the French Republic to General von Manteuffel, Commander-in-Chief of the German Army of Occupation in France, when they praised his impartial, just and humane administration of the occupied French provinces in personal letters to him.

Looked at in this light, the policy of the Washington government seems all the more regrettable, for it is putting an end to General Allen's mission at a moment when its continuance would appear most necessary and when, in deep distress, the Rhenish people have dire need of a counteracting and moderating influence against unbounded political and economic imperialism.

THE DEPARTURE OF THE AMERICANS.

(Translation)

(From "Coblenzer Zeitung", G.P.P., Coblenz, Jan. 24, 1923)

Coblenz entered upon a new epoch of its history at the stroke of twelve today, when the star-spangled banner was hauled down from Ehrenbreitstein.

It was not easy for the citizens of two nations until then wholly unknown to one another to grow accustomed to each other's habits while rubbing shoulders in the narrow confines of the Coblenz bridgehead. When the Americans marched into this city in 1918 they came with the consciousness that their intervention had decided the war. In addition to this, the general tone of the man in the trenches survived to exert its influence during the period of the armistice. There was no end of complaints from individuals as to the harshness of the men. There were also many points of difference in demeanor, and the dissimilarities in the entire mental make-up of the two peoples, which up to that time had been separated by a mighty ocean, led to unavoidable friction. The American soldiers, however, soon realised that the "Boche", who had been so much anathematised in their country amid the smoke and turmoil of war, was not the "Hun" that misguided public opinion in the United States had made him out to be. The American waited a long time for the ill-starred German to reveal himself at last in all his deceit and malice, and then admitted to himself- and to us of the editorial staff - that the German could not conceal his true character for months on end, and that he might after all not be such a deep-dyed villain as the whole world had represented him to be during the war; that as a matter of fact he was, at bottom, quite a likable, genial, industrious, peaceable and good-natured fellow. One factor which helped to bring about better relations was the fact that German blood flowed in the veins of a large, if not a preponderating, number of the American troops of occupation, and that most of the soldiers understood German, spoke German and learned to speak it better. When, furthermore, Lt. Colonel Grier brought about order in the relations between the forces of occupation and the population with a firm hand at the end of 1919, those relations become more and more friendly. Any kind of occupation is a colossal burden pressing down on the population; but under General Allen's command and with the mediation of Colonel Stone and Captain Fieker, the unavoidable friction was reduced to the minimum degree possible. We therefore express merited appreciation of the work of these gentlemen - which appreciation, we feel sure, is shared by the entire population.

We think we may say without fear of contradiction that the private relationship between our own editorial staff and American headquarters was surrounded with an atmosphere of mutual esteem. There was a continuous exchange of news and ideas between us, and information was also exchanged mutually between us and the "Amaroc News". The "Coblenzer Zeitung" always found Headquarters willing to lend an ear to it and give it friendly assistance, and it is our duty as gentlemen to make public acknowledgement of this now that the Americans are leaving.

It goes without saying that the two parties did not always see eye to eye on the matter of political expediency when such exchanges of views took place with members of the headquarters staff. But on the whole there was a considerable community of interest during the four years which have elapsed since Captain Fernbach stated to us at the conclusion of the peace of Versailles that Germany's retention of the Rhineland was due to American intervention. We therefore regret all the more that the American Government's decision to withdraw its troops of occupation, however much it was intended as a protest against the French occupation of the Ruhr, is not in accordance with German interests on the Rhine. It is certain that the inadequate understanding of continental conditions which still prevails at Washington is largely responsible for this act. We thus hope that when the many thousands of Americans have reached their mother-country they will make their better knowledge of the situation felt on matters political, and that America, so badly informed about Germany during the war, will be brought to a full realization of its interests and awake to the fact that Germany's liberty and welfare are as much a necessity to the United States as they are to Europe.

SUB-CHAPTER C.

SANITATION AND PUBLIC HEALTH

General Condition.

The outstanding feature in a review of the records of sanitary conditions in the American area from January 1st, 1922, to January 27th, 1923[1], is the gradual improvement which was brought about in that period. There were no serious epidemics among the civilian population aside from the fairly prevalent influenza during the late winter and spring of 1922. The death rate among civilians was even lower than that among the military.

Births and Deaths.

The civilian birth rate was somewhat lower than that of 1921, but there was little variation in the relative proportion of illegitimate births. The excess of births over deaths was still considerably greater than in the United States and far greater than in France. It was also higher than that of Germany immediately before the war. The following table shows comparative figures on births and deaths in 1921 and 1922:

	1921	1922
Total births	13,484	11,616
Total deaths	6,564	6,448
Percentage of deaths to births	48.68	55.50
Deaths from pulmonary tuberculosis	480	428
Percentage of total death rate	7.313	6.63

[1] A.R. in O.G., 1920-21, Chapter 5.

	1921	1922
Deaths due to influenza	36	66
Illegitimate births	865	669
Percentage of total births	5.838	5.66
Illegitimate American fathers	360	151.2
Percentage of illegitimates	41.62	22.6

Tuberculosis.

Tuberculosis decreased both relatively and absolutely, though it is believed that the coming year may show an increase over the figures for 1922. This disease, which is one of the most serious in Germany, is probably encouraged by the old, unsanitary, damp houses in which a large percentage of the people lives. Improperly heated and poorly ventilated sleeping rooms play a large part, as do also lack of sufficient and proper food and clothing on the part of the great masses of the lower classes. The greatest amount of tuberculosis was found in the cities of Coblenz, Andernach, Neuwied and Mayen. Kreis Ahrweiler, with a population largely rural, reported very few tubercular cases.

Prisons and Hospitals.

The German institutions which were regularly inspected fulfilled the requirements of sanitation. Some of them were more primitive than similar establishments in the United States, but on the whole they corresponded well with the standard of living to which their inmates had been accustomed. The institutions regularly inspected were the Bendorf Frauenanstalt, the Buerger Hospital, Coblenz, the Fischelstrasse jail for women and the Karmeliterstrasse jail. Occasional inspections were also made of other hospitals and charitable institutions in and around Coblenz. Their condition was generally found to be satisfactory.

Hotels and Restaurants.

Inspection of the hotels and restaurants of Coblenz were made at regular intervals. As a rule the conditions were not the best, but it was always possible to obtain at least a temporary improvement by calling obvious defects to the attention of the manager. The most common fault was the failure to furnish kitchen employees and waiters with facilities for washing their hands. This condition was corrected in the hotels and restaurants frequented by Americans.

It was also found that the German restaurants did not have satisfactory ice-boxes. Their typical box was a large, metal-lined container with a lid on the top through which the ice was putin, then a layer of towels or cloths and then the meat, fish, butter or vegetables. It was difficult to wash the box, because such a procedure necessitated taking everything out, including the ice, and consequently towels were often left in boxes longer than a sound sanitary policy would permit, as a result of which occasional cases of ptomaine poisoning might develop.

Barber Shops, Theatres, etc.

Barber shops were regularly inspected, but no serious irregularities were found. Ventilation in theatres and moving picture establishments was usually not up to the American standard, but the houses were uniformly clean.

Food Supply.

The Coblenz slaughter-house was built about twenty-five years ago, yet from a sanitary viewpoint it is satisfactory. The German law provides for what is called a "Freibank", which in the end works out quite well. All meats which have been inspected and passed by the veterinarian bear a number of stamps on the surface, and no other may be sold legally in meat markets or served in restaurants and hotels. The meat of certain diseased animals, which in the opinion of the Veterinarian may be eaten without harm, is sold at reduced prices over this "Freibank" to poorer people who are in possession of tickets entitling them to make such purchases. Not more than four pounds may be sold to one person and sometimes less, depending upon the amount of such meat available and the number of people waiting to be served. Penalties are sufficiently severe to obviate fear of receiving such meat in meat markets or in hotels and restaurants. Swine suffering from contagious pneumonia are slaughtered and the meat cooked and pickled and sold over the counter, the explanation being given that only the lungs are affected. Some horse meat is also sold at Coblenz, as elsewhere in Germany, but mostly only old, worn-out animals are slaughtered for this purpose.

Milk was generally scarce and high priced. This was not due entirely to the delivery of cows to Belgium and France on reparations accounts, but also to the killing of calves for veal, Another reason was the low price paid the farmers for milk. farmers at one time receiving 100 marks per liter for milk, which was re-sold by the middleman for 300 marks.

Improvement in Venereal Conditions.

Ever since the institution of the Provost Court, the women's prison and the Bendorf Frauenanstalt[1], together with the reduction in the number of troops, a gradual decrease had been noted in the amount of venereal disease, but this was especially so in 1922. The number of prostitutes at Coblenz was materially reduced, thanks largely to the activities of the Provost Court and its severe penalties, especially that of deportation. On the date of termination of the American control in this area, the number of women in the Bendorf Frauenanstalt and the Fischelstrasse jail was the lowest in the history of the A.F.G. The improvement since 1921 is shown in the following tables:

Data Concerning Bendorf Frauenanstalt.

	1921	1922
Average number of inmates	244.7	156.
Average with gonorrhoea	155.0	87.8
Average with syphilis	48.0	35.0
Average with gonorrhoea and syphilis	42.0	30.0
Chancroid, observation, babies, etc.	9.0	4.2

Provost Court records show conclusively the decrease in the number of women arrested on statutory charges and the lower number of women infected among those arrested in 1922 as compared with 1921:

---

[1] The administration of the Frauenanstalt, under Dr. Kurt Roscher, during the past three years deserves special mention. Dr. Roscher is one of the foremost authorities on venereal diseases in Germany. This, combined with his interest in the work and his humanitarian treatment of his patients, a great many of whom were of the very lowest type, made of the Frauenanstalt an institution of inestimable value to the American Forces in Germany and to the German population of the American occupied area.

	1921	1922
Total Arrests	2554.0	1413.0
Average per month	213.0	117.8
Total convictions	2237.0	1280.0
Average per month	186.4	106.6
Total infected	799.0	453.0
Average per month	65.6	38.0

SUB-CHAPTER D

PUBLIC UTILITIES

Functioning of Utilities.

The existing arrangements in regard to public utilities[1] continued in effect throughout the year 1922 and up to the date on which the American area was handed over to the French. In general, little or no disturbance in the delivery of these utilities was noted by the ultimate consumer, although circumstances which might have led to confusion were met and disposed of by the Engineer Officer of the A.F.G. Coal shortage, labor troubles and inadequate provisions for water supply were factors of worry to Headquarters, but at no time was the situation so serious that the troops were put to any inconvenience as a result thereof.

Difficulties Encountered in Delivery.

The principal difficulties during 1922 were caused by labor disagreements. A tie-up of rail transportation for about fifteen days in February ensued as a result of the locomotive drivers' strike. While this did not generally affect the utilities in the area, since the minimum reserves of coal required were maintained intact with but few exceptions, it caused a fuel shortage at the

[1]See A.R. in O.G., Vol. II, Chapter 6.

Concordia Huette, at Bendorf, which supplies about 75% of the gas for Coblenz and near-by towns. The I.A.R.H.C. and the D.G.C.R.A. thereupon took action, their investigations showing that the non-deliveries to the American occupied area were caused not only by the rail tie-up but also by the Ruhr coal deliveries to France for reparations. The minimum reserve of fuel at the Concordia Huette was again attained only after a period of several months, during which time Coblenz and the vicinity were restricted to 75% of the normal consumption of gas.

The Hoehn Power Plant Strike.

Two short-lived strikes occurred at the Hoehn power plant in unoccupied territory, thus affecting the distribution of the necessary current to the Coblenz electric light and power plant. The reason given by the workers for the strike was that they wished to have some days off in which to work in their fields, most of them being farmers.

The Walk-outs at Concordia Huette.

The account of the walk-out at the Concordia Huette referred to in sub-chapter A[1] is worthy of repetition because of the prominent part played therein by the American authorities. The blast furnaces and coking plant at Bendorf were classified as a public utility and came under the I.A.R.H.C. ordinances regulating strikes, inasmuch as a shut-down of either unit affected Coblenz and the

---

1 see page 225.

adjoining towns, they being dependent upon the Concordia Huette for a certain supply of gas produced as a by-product of the coke ovens. As the lay-out of the works did not provide for storage of large quantities of coke, the coking plant could not be operated for more than three days unless the blast furnaces were kept in operation to consume the coke. On March 18th, however, the workers at both the coke ovens and blast furnaces left their places. It was evident that the leaders had either lost control of the strike situation or were deliberately testing the patience of the occupying forces. Detailed plans were prepared in the office of the Engineer, A.F.G. for requisitioning the plant and necessary labor to operate it in case work was not resumed. The workers were finally convinced that the American authorities, represented by a member of the I.A.R.H.C., would tolerate no disobedience of their orders to resume work, and operation of the ovens and furnaces was resumed. Two days later an agreement with the employers was reached and the strike was ended.

Another walk-out occurred on March 27th, and again the public utilities section of the works was affected. This time action was more drastic and proceedings were undertaken with a view to deporting the agitators responsible for the trouble. Work was resumed after a few hours.

On July 15th, without warning and in violation of I.A.R.H.C. ordinances, another walk-out caused the works to close down. Orders were issued by the Americans to the labor leaders to provide the necessary personnel for the operation of the coke ovens and blast furnaces. These orders not being obeyed, that portion of the works was requisitioned, information to this effect being furnished the plant director and the strike leaders and orders given them to provide the necessary laborers. In addition, the American authorities rushed two officers and thirty-two men, together with such Military Police as were available, to the works for guard duty. This show of force, with the requisitioning of the plant, had the desired effect and all the laborers on the night shift reported for work with only few exceptions. Some strike leaders were arrested, tried by the American Provost Court and sentenced to confinement or fines, and others were deported from the area.

Improvements in Water Supply.

The heavy rains and snows of the winter of 1921-22 and the spring of the latter year effectively broke the two years' drought which had caused so much difficulty in various communities dependent upon springs for their water supply. Notwithstanding the fact that the output of all the springs had returned to normal, the first warm weather in May caused a water shortage at Pfaffendorf, so that a certain amount of water had to be diverted from the Ehrenbreitstein mains. In this particular instance the German authorities could not place the blame upon the unusual amount of water required by troops, since the town was unoccupied. This proved that the fault lay in the inadequate provisions made for the increase in consumption and the decrease in the output of the springs with time. A six-inch water main crossing the Horchheim bridge was completed in August. This connected the Coblenz water works with the water supply systems of Horchheim, Pfaffendorf and Ehrenbreitstein and relieved those towns from absolute dependency upon springs for their supply.

Effect of Ruhr Occupation.

From January 18th, 1923, to the date on which American control in the Coblenz bridgehead ceased, no coal was received by any public utility, this being a development of the occupation of the Ruhr by the French. Consequently, all reserve stocks of fuel were drawn on heavily. The shortage at the Hoehn electric plant necessitated continuous operation of reserve generators at the Coblenz plant. Gas supply was maintained at normal, though the length of time for which such could be assured was uncertain.

The non-delivery of the amount of fuel required for the utilities of the American area had brought about a serious situation on January 27th, and it was feared that the production of gas and electric plants would be curtailed as a result. While steps were taken to ensure deliveries of the necessary fuel, no definite assurance had been received from the Interallied Military Coal Commission at Essen that it would be forthcoming.

SUB-CHAPTER E.

WELFARE WORK

The Hoover Commission Child Relief work begun in 1920[1] was continued on a limited scale throughout 1922 and, not being an Army activity, was projected to extend beyond the time at which the American troops would evacuate the Rhineland. 203, 630 meals were served to under-nourished children in the following localities in the American area, as well as in other Rhineland cities and towns where much procedure was necessary, in 1922:

> Coblenz City; Coblenz Land; including three orphan asylums; Kreis Mayen; Kreis Wirges; Neuwied; Niedermendig; Herachbach.

The members of the American Forces in Germany, following the custom practiced during the preceding three years, organized charity activities for Christmas of 1922. Whereas there were approximately 10,000 troops in the American area at Christmas, 1921, by whom the sum of $4,715.00 was contributed for the benefit of needy German children, the strength of the forces in 1922 was somewhat less than 1,200. This small body gave voluntary donations amounting to $1,500.00, most of which was expended for articles of wearing apparel, such as stockings, underwear, sweaters, shoes, etc., for the most needy children.

There were but three centers of distribution, namely: Coblenz proper, Coblenz-Luetzel and Ehrenbreitstein (all troops had been withdrawn by this time from the outlying towns of the American area). In each of the three places mentioned a

---

[1]See A.R. in O.G., 1920-21, Vol. II , pp. 97 to 99.

separate committee managed the local situation under the supervision of a central committee. Approximately 1,500 children were given presents by the three committees, 800 more were given special dinners and a number of poverty-stricken families received contributions of foodstuffs. After all bills incurred in the purchase of presents had been paid, there remained, due to a sudden fluctuation in the exchange before the dollars collected were converted into marks, a sum of approximately 1,250,000 marks, which was distributed among the German charitable institutions for the aid of the most deserving families during the winter.

The Christmas spirit was further demonstrated to a nicety by the local posts of the American Legion and the Veterans of Foreign Wars, both of which were hosts at dinners for several hundred of the poorest children of Coblenz.

SUB-CHAPTER F.

FINANCE.

General Policy.

The policy of the Finance Department of the A.F.G. did not change materially from that followed during the latter part of 1921.[1]

Costs of Occupation.

The cost of the American Army of Occupation for the calendar year 1922 was $4,662,733.48, or an average monthly cost of $388,561.12 all of which was chargeable to Germany. From January 1st to June 30th, the cost was $3,501,891.96, and from July 1st to December 31st, $1,160,841.52. The cost of the second six months was affected by the return of the major portion of the troops to the United States before June 30th.

Procurement of Funds.

German paper marks were requisitioned from the German government as needed. During the year, 865,681,561.45 paper marks (armistice funds) were received and disbursed at a daily rate of exchange furnished by the local representative of the German government per the agreement put into effect on October 26, 1921. The dollar equivalent

---

[1] See A.R. in O.G. 1920-21, Vol. II, Chapter 9.

of the marks received and disbursed was $760,865.80, or approximately 16-1/3% of the total cost of the Army of Occupation. All of these marks were received and disbursed without any loss to the United States on account of fluctuation and depreciation.

All demands made upon the German government for marks were complied with immediately, within the limit of the authorised requisitions, through the Reichsbank at Coblenz. Relations between the Finance Officer and the agencies of the German government entrusted with the settlement of requisitions for funds were at all times most satisfactory.

Disbursements.

Only the Coblenz Disbursing Office continued in operation throughout the entire year, that in Antwerp being closed on June 30th, 1922. Payments were made at the former office to commissioned officers, warrant officers, field clerks and nurses, either in U.S. currency, Treasury checks or in German marks, at the option of the persons concerned, and to enlisted men in U.S. currency, with the option of exchanging any part or all of their pay for German marks at the time of payment. The total disbursements for the calendar year were $2,959,455.80.

German Payment for Army Utilities.

In addition to the amount of paper marks furnished the Finance Officer upon requisition, the German government paid for all German labor employed by the different services of the Army of Occupation, for all billets, for forage, fuel, light, heat, gasoline, oil, transportation in the occupied area and, by a protocol entered into on August 11th, 1922,[1] for transportation of all personnel to the port of debarkation in New York and for transportation of supplies, etc. from New York to Coblenz.[2] This effected a large saving in appropriations for the support of the Army during the year.

Up to June 30th, 1922, no commutation of quarters or heat and light for officers or others serving with these forces was paid, except for dependents living outside of the occupied area, and no rental allowance since July 1st, 1922. Had these forces been stationed within the territorial limits of the United States they would have been furnished either quarters in kind at government expense, including maintenance of buildings, fuel, light, water, heavy furniture, etc., or commutation in lieu of quarters if no public quarters had been available. The act of June 10th, 1922, eliminated foreign service pay of 10% and 20%, so that no additional cost was incurred on that account for the maintenance of the forces on the Rhine after July 1st, 1922.

Currencies Dealt in.

Foreign currency received from sales of surplus

---

[1] See page 369.

[2] See Chapter 4, sub-chapter F, Appendix 1.

supplies, salvaged material, Provost Court fines, etc., amounted to: German marks 113,716,751.86, French francs 940,467.35, Belgian francs 469,996.57. During the year, marks to the value of $235,662.79 were sold to local banks at the official rate of exchange, or better. French francs were either disbursed in the American area or were transferred to the Finance Officer at Paris for disbursement by him. Belgian francs were disbursed either at Coblenz or by the Finance Officer at Antwerp, as long as that office continued, or were still on deposit in the Antwerp branch of the National City Bank of New York.

Depreciation of Mark.

The German mark depreciated rapidly during 1922. The lowest rate of exchange between the mark and the dollar was 170.94 marks for one dollar in January, and the highest 8,421.05 marks for one dollar in December. The mark continued to fall after the beginning of the New Year, reaching its lowest point during the American stay on the Rhine on January 27th, 1923, viz., 23,809.52 marks for one dollar. When the rate of exchange was favorable, officers and soldiers took marks for their pay, but when the commercial rate was better than the government rate, payment was taken in U. S. currency.

Property Auditing and Cost Accounting.

Owing to the reduction of the commissioned personnel of the Finance Office from five officers to two, it became necessary on June 30th for the Finance Officer to take over from his assistants the duties of property auditing and cost accounting, the remaining officer being the Disbursing officer of the A.F.G. on January 1st, 1922, the personnel of the office of the Auditor, A.F.G., had consisted of one commissioned officer, as assistant to the Finance Officer, four enlisted men and two civilians (one American and one German). This force was gradually reduced until the end of the year, when only one enlisted man and two civilians remained to perform the above duties under the jurisdiction of the Finance Officer.

Cost reports on the American Army of Occupation were compiled from data furnished by the various services and were submitted quarterly to the Reparations Commission at Paris.[1] These reports also included the expenses and receipts of the American section, Interallied Rhineland High Commission.

An audit was made of all property accounts of the A.F.G. during the year, many of them having been closed out due to the discontinuance and disbandment of a great number of temporary organizations and the return of a large percentage of the personnel to the United States. It was the duty of the Auditor, A.F.G., to see that property shortages were settled by the responsible officers through use of survey or some other competent voucher or by payment thereof to the Finance Officer. In addition, all officers relieved from duty with the A.F.G. were required

---

[1]See A.R. in O.G., 1920-21, Vol. II, pages 122-24.

to obtain a final clearance, covering money and property accountability, a clearance from billets, private debts, etc. from the Auditor and, after June 30th, from the Finance Officer.

Officers.

The following officers were on duty in the Finance Officer for the periods indicated:

Lieut. Col. L. A. Sturges, F.D.,	Jan.1,1922 - Jan.27,1923.
Major E. E. Lockard, F.D.,	Jan.1,1922 - Dec.31,1922.
Major G. R. Doane, F.D.,	Jan.1,1922 - June 20,1922.
Major H. G. Rice, F.D.,	Jan.1,1922 - May 13,1922.
1st Lieut. J.H.McFall, F.D.,	Jan.1,1922 - June 30,1922.

SUB-CHAPTER G.

RENTS, REQUISITIONS AND CLAIMS SERVICE.

Activities.

The Rents, Requisitions and Claims Service, A.F.G., continued its functions throughout the year 1922 and was still in operation on the day that the American area was turned over to the French. Arrangements were made for its representative to remain as a member of the "clean-up squad" to insure investigation and payment of all claims which might be presented after the bulk of the forces had left.

At noon on January 27th, 1923, the R.R. & C. Service discontinued requisitions for the I.A.R.H.C., the Interallied Railway Commission (French) and the French Army. The Local Assessment Commission ceased operating on January 30th.

The R.R.& C. offices in the Second Brigade area were closed on the withdrawal of that organization in the spring of 1922. French troops replaced the Americans in the area, the same policy continuing in effect for the French Army in the American area as had existed for the American Army. The R.R.& C. office at the Port of Antwerp, Belgium, was closed on June 30th, 1922, unsettled business being acted upon by the Coblenz office.

Rents.

The Billeting Commission, the composition of which is given on page 386, A.R. in O.G., 1920-21, met monthly to consider questions pertaining to the billeting of the armies of occupation in the American area. At these meetings, rentals for billets occupied by members of the armies of occupation and the Interallied Rhineland High Commission were submitted by the German Billeting Commission, composed of representatives of the German government and city officials. Before these rentals were put into effect, they were checked by the R.R. & C. Service and approved by the Billeting Commission.

The following compilation shows the amounts paid by the German government for rentals during 1922:

For single billets -	Jan. 1 to June 30, 1922 -	2,425, 661,77	Mks.
	July 1 to Dec. 31, 1922 -	1,517,321.00	"
For quarters and land -	Jan. 1 to June 30, 1922 -	16,220,660.09	"
	July 1 to Dec. 31, 1922 -	41,272,995.08	"
For furniture -	Jan. 1 to June 30, 1922 -	1,365,518.12	"
	July 1 to Dec. 31, 1922 -	4,576,567.65	"
For repairs -	Jan. 1 to June 30, 1922 -	501,117.54	"
	July 1 to Dec. 31, 1922 -	8,158,123.88	"
Value of furniture and furnishings returned to central German government to be credited to costs of occupation -	Jan. 1 to June 30, 1922 -	6,155,894.68	"
	July 1 to Dec. 31, 1922 -	4,195,881.60	"

Requisitions.

Requisitions for the American Army, Interallied Rhineland High Commission and the French Army in the American area were all dealt with by the Rents, Requisitions and Claims Service in order that proper supervision and control might be exercised by the Commanding General.

The following compilation shows the number of requisitions effected during the year for the three bodies mentioned:

<u>Real property (A.F.G. and I.A.R.H.G.O:</u>	
In effect December 31, 1921	921
In effect June 30, 1922	520
In effect December 31, 1922	477
<u>Furniture and repairs (A.F.G. and I.A.R.H.C.):</u>	
Issued to December 31, 1921	3742
Issued to June 30, 1922	5055
Issued to December 31, 1922	6343
<u>Real property (French Army):</u>	
In effect December 31, 1921	0
In effect June 30, 1922	63
In effect December 31, 1922	267
<u>Furniture and repairs (French Army):</u>	
Issued to December 31, 1921	0
Issued to June 30, 1922	39
Issued to December 31, 1922	1096

On January 27th, 1923, 170 A.F.G. requisitions continued in effect, also 288 I.A.R.H.C. requisitions and 269 French Army requisitions, making a grand total of 727 requisitions in operation in the American area on January 27th, 1923, when control was handed over to the French.

<u>Claims.</u>

The Commanding General directed that the Local Assessment Commission, established by Ordinance 59, I.A.R.H.C., begin functioning in the spring of 1922. This commission, whose duty was to determine the charges for requisitions effected in each zone and estimate the damages caused by the troops of occupation therein, held its first meeting on May 19th, 1922. The Local Assessment Commission in the American area assessed all claims arising subsequent to November 11th, 1921, the date of entry into effect of the treaty of peace between the United States of American and the German Republic. Claims were submitted by German nationals through their local burgomasters, investigated by the local Town Major and then forwarded to the Rents, Requisitions and Claims Service with the recommendation of the Town Major. Those in excess of 5,000 marks were re-investigated by the Investigating Officer of the R. R. and C. Service, after which they were forwarded to the Local Assessment Commission for action. Claims for less than 5,000 marks were forwarded to the Local Assessment Commission without further investigation. This decision of the Local Assessment Commission was without appeal, notices being forwarded to the claimant and to the Reichsvermo egensverwaltung (German Property Office) of the findings of the Local Assessment Commission on each claim, except in a small number of instances in which the damage was not due or incident to the Army of Occupation and the 105th Article of War was not applicable. Claims which had been assessed by a board of officers under the 105th Article of War were considered closed upon the approval of the reviewing authority of the board and were not forwarded to the Local Assessment Commission.

The Interallied Rhineland High Commission and the French Army in the American area were authorized to investigate direct claims submitted by German nationals against the organizations mentioned. These claims were forwarded after investigation to the Local Assessment Commission by the parties concerned.

The following compilation shows the status of claims submitted by German nationals against the American Army of Occupation and its members:

Claims on file:

	Dec. 31, 1921	June 30, 1922	Dec. 31, 1922	Jan. 27 1923
Approved -	26408	26928	27416	27516
Disapproved -	24544	24816	25246	25280
Total --	50952	51744	52662	52796

Unsettled claims on file - December 31, 1922 -- -- 135
January 27, 1923 -- -- 38

Billeting.

The number of individual billets occupied decreased substantially, due to the continuous withdrawal of the forces. Only a few billets remained occupied on the day that the 8th Infantry was returned to the States, as will be seen by the following table:

Individual Billets (Zone Major's office);

Billets occupied on January 1st, 1923 . .	233
Assigned during month	12
Vacated during month	96
Occupied on January 27, 1923.	149

Officers.

Officers on duty with the R. R. and C. Service for all or part of the period covered, were:

Captain C. C. Bodeker, Infantry.
Captain R. L. Creed, Cavalry.
Captain Oscar K. Wolber, Infantry.
Captain M. A. S. King, F. A.
Captain C. R. Breckenridge, Jr., Infantry.
1st Lieut. Lloyd M. Hanna, C.A.C.

SUB-CHAPTER H.

DISPOSITION OF SURPLUS STOCKS AND ANIMALS.

Necessity for Sales.

In continuation of the policy adopted in August 1, 1920,[1] the disposition and sale of surplus stocks and animals remained a Quartermaster activity. Due to the numerous orders received for the return of units and individuals from these forces, the accumulation of surplus stocks was hard to avoid, as returning units, especially, turned in practically all their heavy equipment and, in most cases, all their animals. While this served to make easier the question of supply for those remaining, it nevertheless made a strenuous sales campaign necessary.

Details of Sales.

A total of $879,481.69 was received by the General Sales Agent from sales of government property, including Quartermaster, Ordnance, Medical and Engineer supplies and horses and mules in the year 1922. In addition, property valued at $606,191.99 was transferred from the Army stocks to the Department of Agriculture, this comprising Quartermaster property estimated at $458,804.91, Engineer property at $92,325.27 and Ordnance property at $55,061.81.

Seven hundred and seventy-seven mules were sold at an average price of $158.00 each, 462 light draft horses at an aver-

[1] A.R. in Q.G., 1920-21, Vol. II, p. 189.

age price of $143.00, and 310 riding horses at an average price of $125.00. All these prices were considered good and above medium, for many of the animals were old and in poor condition. The methods of selling were public auction, sealed proposals, and negotiations.

The French Army contracted for the purchase of TNT, fuses, cartridges, 75 mm. explosive shells, grenades, targets and target paper to the amount of $31,760.27 on May 15th. Sealed proposals on a lot of ammunition were opened on June 13, the highest bid - the equivalent in marks of $10,558.60 - being submitted by the Eisen Industrie of Essen. This being war material, it was necessary for the purchaser to break it down before its removal from the occupied area was permitted.

All Air Service activities ceased in April, whereupon the entire lot of aviation material at the Weissenthurm Flying Field was advertised and sold on sealed bids.

The Ryan Corporation Contract.

The biggest private contract entered into during the year was that between the American Army, on the one part, and the Ryan Corporation, on the second part, for the purchase, at the prices listed, by the party of the second part, of the following surplus Motor Transport property: pneumatic tires, $74,840.00; inner tubes, $13,000.00; solid tires, $26,090.00; automobile spare parts and accessories, $57,000.00; Motor Transport shop equipment, $40,000.00; and trucks, trailers and motorcycles, $60,070.00. Shortly afterwards, orders were received for the retention of a small force on the Rhine, and at the same time instructions to the effect that the Department of Agriculture be permitted to select such road building material as desired. The amount of supplies which the Ryan Corporation was to receive was therefore considerably reduced. On the date on which General Allen gave over control of the area to the French, the Ryan Corporation had taken delivery of a total of $47,855.38 worth of Motor Transport property.

The Final "Clean-up".

Plans for the disposal of all surplus stocks left by the Army on its actual evacuation of the Rhineland had not been announced by January 27th, 1923. The goods to be sold consisted of: motor vehicles, horses, mules, subsistence stores, paints, oils, and greases. It was known, however, that the methods to be used would be sale under sealed proposals and at public auction, the latter, in all probability, for horses and mules.

Officers.

The following officers were on duty throughout the year at the places and for the periods indicated:

Lt. Col. Henry H. Sheen, Q.M.C., General Sales Agent from January 1, 1922, to date of final disposition of all property.

Maj. R. W. Riefkohl, Q.M.C., Assistant General Sales Agent, Paris, January 1, 1922-August 31, 1922.

Maj. C. L. Corbin, Q.M.C., Assistant General Sales Agent, A.F.G., January 1, 1922 - August 30, 1922.

Capt. H. W. Beyette, Q.M.C., Assistant General Sales Agent, Paris, August 31, 1922 - October 13, 1922.

Capt. E. P. Doyle, Q.M.C., Assistant General Sales Agent, Paris, October 13, 1922, to date of final disposition of all property.

Captain E. M. Scott, Q.M.C., Assistant General Sales Agent, A.F.G., January 1, 1922 - August 30, 1922.

Captain John Biggar, Q.M.C., Assistant General Sales Agent, Antwerp, March 30, 1922 - June 30, 1922.

1st Lt. G. F. Chapman, Q.M.C., Assistant General Sales Agent, A.F.G., September 2, 1922, to date of final disposition of all property.

SUB-CHAPTER I.

DISMANTLEMENT OF FORTIFICATIONS.

Destruction of Fort Kaiser Alexander.

The dismantlement of German fortifications in the American occupied area[1] continued during 1922 under the provisions of Article 180 of the Versailles treaty, which reads in part as follows:

> "Those which are situated in territory occupied by Allied and Associated troops shall be disarmed and dismantled within such periods as may be fixed by the Allied High Command".

The Interallied Military Control Commission, with General Nollet as its President and Headquarters at Berlin, in the supervision of that part of the treaty of Versailles quoted, issued instructions on February 11th, 1922, for the immediate dismantlement of Fort Kaiser Alexander. This fort, which was considered one of the greatest fortifications in existence upon its completion in 1822 and after the addition, in the years 1865 to 1870, of exterior walls and ramparts, was of value only as a defensive element, being provided with overhead cover of approximately fifteen feet in various places. Destruction of the subterranean passage connecting Fort Alexander and Fort Constantin was undertaken in addition, it being classified in the category of those works to be partially dismantled.

Status of Fort Ehrenbreitstein.

---

[1]See Chapter 13, Vol. II, A.R. in O.G., 1920-21, for details of previous dismantlements.

Fort Ehrenbreitstein remained untouched, continuing as the garrison for the machine gun companies of the 8th Infantry. Marshal Foch, in a letter to the Commanding General on February 9th, recognized and agreed to General Allen's contention that it should not be destroyed but preserved because of its historic significance as one of the famous monuments of the Rhine.

Details of Demolition.

Contracts for dismantling under the supervision of the Entfestigungsamt (Dismantling Office of the German government, the personnel of which was composed of former German officers), were lst on March 13th to firms of Essen and Mannheim. Work began on April 10th, the intervening time being spent by the contractors in securing, transporting and placing equipment. Dismantling by demolition was the first step. After some progress therein, equipment for the removal of debris and earth was installed and put in operation. The work of dismantling was completed in December.

Explosives for dismantling purposes were imported from unoccupied territory under Interallied Rhineland High Commission regulations governing import, transport and use of such materials, and were stored at Fests Franz, Fort Constantin and Fort Rheinhell. The following quantities were used to demolish the works mentioned above:

154,000 pounds of explosive
3,700 metres of time fuse
3,500 caps, electric
3,500 fuse lighters

In the removal of debris and earth works the following equipment was used:

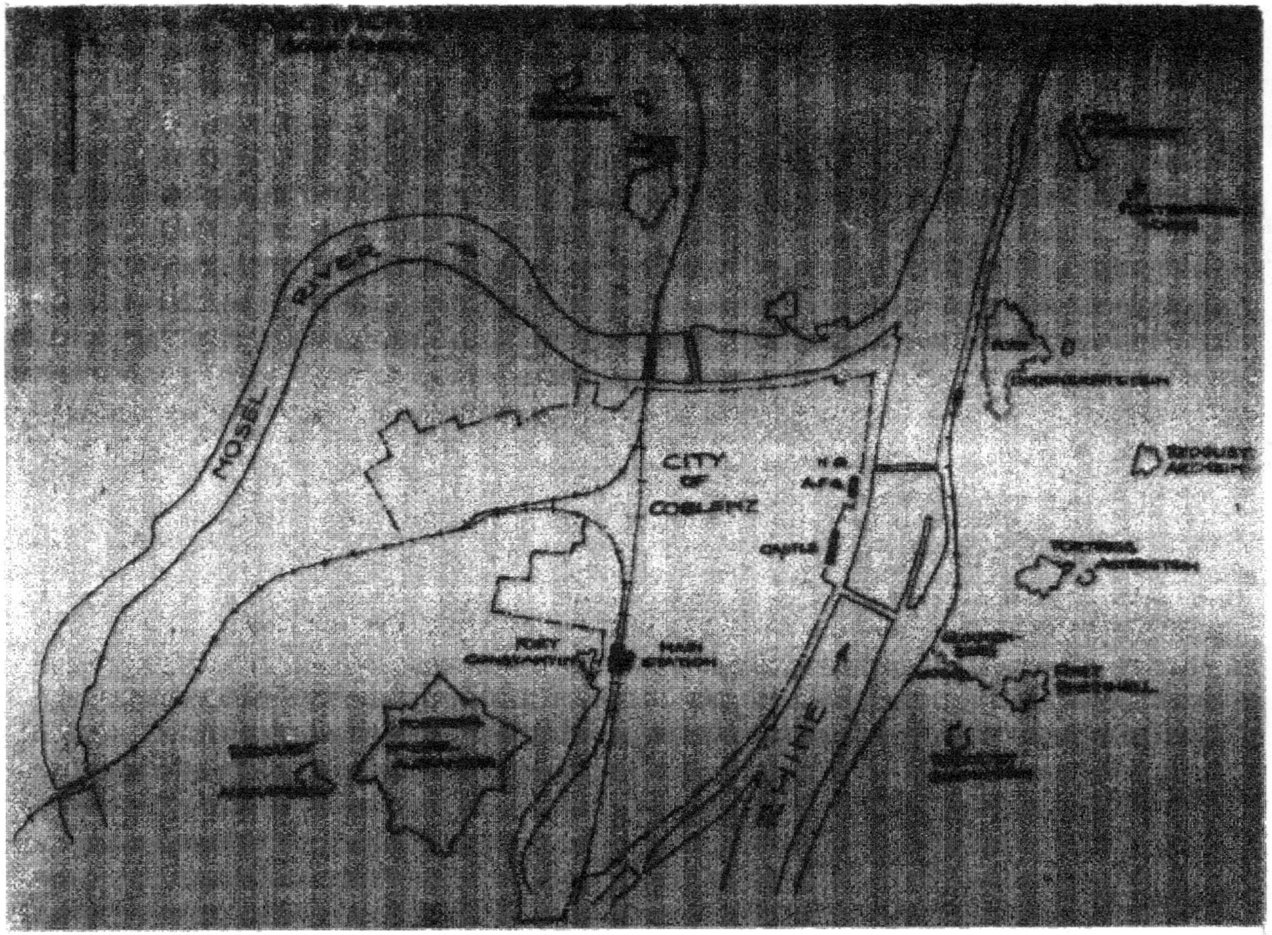

MOSEL RIVER
CITY OF COBLENZ
RHINE
CASTLE

1 steam shovel or dredge
6 locomotives
100 dump cars
5,500 metres of 60 cm. railway track

Personnel employed in dismantling was as follows:

Month	Engineers	Foremen	Laborers
April	1	5	33
May	1	3	84
June	1	4	201
July	1	9	366
August	1	15	452
September	1	9	259
October	1	8	260
November	1	6	242

The total of materials destroyed and removed was:

140,000 cubic metres stone, of which 16,000 cubic metres were salvaged for building construction:

130,000 cubic metres earth.

Due to the fluctuation of German currency, the cost of dismantling and disarming Fort Kaiser Alexander was far in excess of the estimates, amounting to approximately 35,000,000 marks.

Officers.

Major Edwin A. Bethel, C. of E., had supervision of the dismantling of fortifications in the American area until December 10th, 1922, when, due to his return to the States, the duty was taken over by First Lieut. J.H. Harman, C. of E.

SUB-CHAPTER K

INTER ALLIED COMMISSIONS.

Interallied Railway Commission.

This body operated as in the preceding year, the personnel being reduced with that of all the other elements of the A.F.G. While the number of control posts was cut down condiserably, efficient supervision over all shipments of goods was still maintained.

The new year of 1922 opened with a strike of the lower grade railway employees and laborers. In this strike, the unions absolutely disregarded the Interallied prohibition to interfere with traffic in the occupied territories. By requisitioning labor and arresting some of the radical leaders, the Allied authorities brought an end to the strike early in the month. In reality, the return to work was due more to orders of the unions for the men to resume work than to any measures taken by the Allied military and civil authorities.

The end of January brought on another strike, which was general throughout Germany with the exception of the occupied areas, where orders issued jointly by the Commanding General, Allied Armies of Occupation, and the Commanding General, A.F.G., had the effect of preserving normal conditions. This evidence of a firm and uniform attitude on the part of the joint authorities resulted in a complete absence of disorders and in practically no suspension of traffic in the occupied area. The grant of a 30% increase in wages to railway officials and laborers shortly thereafter served to bring about a more satisfactory spirit on the part of the railway personnel. At the same time, railway passenger and freight rates were increased accordingly.

No strikes or disturbances of any kind marred the second quarter of the year. Train schedules were improved, the service was more satisfactory, and several new express trains were put on the Rhineland lines. Another 20% increase in pay, together with a corresponding rise in transportation rates, came into effect on June 1st.

The second quarter was characterised by many train movements between Coblenz and Antwerp. These were caused by the transfer to the states of personnel and matériel from these forces in accordance with the plan for the reduction of the A.F.G. and consisted of special trains carrying individuals and units, and solid trains of supplies, army stocks, horses, etc. In addition, numerous movements of men and horses to horse shows and race meetings held by the Allied armies of occupation were made, all without obstruction or casualty.

On June 30th, 1922, the American control of transportation in the bridgehead continued as in the preceding two years, with only the natural reductions of control posts made necessary by the diminution of effectives. Although the 16th Railroad Engineers remained as part of the A.F.G., the decrease in strength of the Quartermaster personnel had necessitated the employment of men of the 16th Engineers for clerical work connected with the Transportation Service, thus making necessary a revision in the strike plan concerning the American area for the future.

Since the withdrawal of the Third Army in 1919, the American Army of Occupation had at its disposal twenty-five refrigerator cars and ten tank cars, formerly belonging to the A.E.F. but left at the disposal of the Americans by the French government, although they had been sold and allocated to various French railway companies. It having been impossible to reach a satisfactory agreement concerning the price to be paid for the rental of these cars, they were all returned to the French government in April and May and a final settlement for them was made. After this time, such cars of this nature as were needed by the A.F.G. were obtained from the German railways in the same manner that other rolling stock was obtained.

No serious strikes occurred during the second half of the year or up to the moment when the American authorities turned over the area to the French. The only strike worthy of note was that of shopmen at Ludwigshafen in December. These men returned to work after the German railways had partially acceded to their demands.

The Line of Communication through Bremen, as

instituted by the protocol of August, 1922, was eminently satisfactory.

Interallied Waterways Commission.

The high water level of the Rhine kept river traffic at a maximum throughout the year. Much extra tonnage was carried by barge during the first quarter due to the labor troubles on the railways, which prevented the carriage of freight by land. Water transport continued above average throughout the entire year because of the resumption of commercial activities by Rhineland factories in consequence of the fall of the mark, which kept all factories working overtime.

Labor troubles on the river were practically unknown, due to judicious increases in wages whenever disturbances threatened.

D. G. C. R. A.

This organisation (Direction Général de Communications et Ravitaillement d'Armée - Headquarters Army Communications and Supplies) continued the functions[1] undertaken in 1918. Two meetings were held daily during the first railway strike of January, 1922, in the offices of the General, D.G.C.R.A., at which representatives of all the armies on the Rhine were present. The latest information on developments was imparted and decisions were made accordingly. Both during and after this strike a lack

---

[1]See A.R. in O.G., 1920-21, vol. II, page 222.

of uniformity was evident in the ideas and policies regarding the methods to be pursued in the furute to prevent such strikes.

While in the midst of a search for proper and uniform remedial measures, the D.G.C.R.A. was confronted by the second January strike. Meetings similar to those convened early in the month were held daily in the office of the General, D.G.C.R.A., and constant telephonic connection was maintained between the American Mission and the A.C.of S., G-4, the Chief of Staff, A.F.G., the American representative on the I.A.R.W.C., and, in several cases, the Commanding General, A.F.G. At the conclusion of this strike in the first week of February, the Commanding General, Allied Forces of Occupation, and the Commanding General, A.F.G., each saw fit to express to the railway personnel that had remained at work during the strike their appreciation of the workers' orderly attitude and their maintenance of the service.

Due to the definite organisation of the Saar Basin by the Saar Government Commission, it became necessary in April to effect a separate agreement with the Commission concerning traffic of each one of the Allied armies through that territory. Negotiations were well under way for an agreement concerning the A.F.G. when orders were received directing the complete withdrawal of the Americans. Upon receipt of these orders the negotiations, which were being conducted through the American Mission and the General,

D.G.C.R.A., were mutually dropped. They were not resumed, due to the extremely low strength of the A.F.G. and the infrequent traffic of members thereof across the Saar territory.

It having been definitely decided by the War Department to discontinue the American base at Antwerp on June 30th, 1922, negotiations were opened between the Commanding General, A.F.G., and the German federal government to obtain space and facilities in a German base port for the small quantity of traffic necessary to supply the A.F.G. The transportation phase of these negotiations was conducted jointly by the Office of the A.C.of S., G-4, the D.G.C.R.A and the German Railway Administration.

In order "to insure steady and uninterrupted railway traffic in the Rhineland," the General, D.G.C.R.A., projected an extensive plan of identity cards for all railway employees in December. The German railway authorities objected to the execution of such a plan as being contrary to the provisions of the I.A.R.H.C., whereupon the extent of the plan was limited to railway officials, laborers being exempted.

In October, General Allen protested against a decision of the Ambassadors' Council authorising Marshal Foch, through General Payot, D.G.C.R.A., to take necessary steps toward the destruction or limitation of various stretches of railway in the Rhineland supposed to be strategic. The General based his protest on the ground that the I.A.R.H.C. was the supreme representative of the Allied and

Associated Powers in the Rhineland, and that therefore the Council of Ambassadors was not competent to give such instructions. An attempt was made in a subsequent meeting of the High Commission to prevent General Allen's letter bearing on the destruction from going forward with the Commission's letter to the Council of Ambassadors. The British and Belgian commissioners, however, approved it, and the opposition was overcome.

Supply.

The D.G.C.R.A. had as its main duties the functions of communications, supply and liaison. The subject of communications has already been covered under the report on the I.A.R.C. and the I.A.N.C.; the Roads Commission and the Interallied Telegraph Bureau, as stated in A.R. in O.G., 1920-21, were not at all permanent, and met only when the situation required. Practically all the functions of liaison were eliminated or passed into the background during the year 1922, with the exception of the plan of defence, which is treated in this history as a function of G-3.

The question of supply continued to be an ever-important factor. The coal supply of the American area was almost depleted early in the year as a result of the railway strike. The weather during the strike period was particularly severe, and much damage was done to the idle locomotives through frozen engines and water tanks. On the conclusion of the strike, energetic efforts were made by the D.G.C.R.A. to provide for a thorough and efficient stocking up of the Army zones during the spring and summer, before time for the moving of the crops. This had always been the policy of the American authorities. By the end of the year, all fuel stocks of utilities had been brought up to a thirty-five day reserve, except at the Concordia Huette. The condition of this utility was receiving the attention of the Coal Control Committee of the I.A.R.H.C. at the time American control of the area was given up.

At the monthly conferences of January, February and March, 1922, the matter of uniform prices for supplies furnished the four Armies of Occupation by the German government was taken up. The matter was discussed exhaustively and progressed to the extent that a comparative tabular statement was prepared showing the prices paid for requisitioned supplies in the four armies. This was necessary, due to the fact that the supreme Allied authorities and the Reparations Commission had set a monetary limit which the Armies of Occupation should not exceed per year. In this respect, the policy of the American Army authorities had always been to determine as nearly as possible the actual cost to the Germans of the supplies furnished. In other words, the idea was not to permit the Germans to over-charge the Army and, on the other hand, not to burden Germany with unnecessarily heavy expenses.

The supply problem for the American Army was simplified by the reduction of the A.F.G., the surplus supplies, on the withdrawal of continents, being used for the daily needs of those remaining. After the discontinuance of the Army Transport service, local sources of supply were drawn on more heavily. After June, 1922, all supplies coming from the United States for the remaining troops were brought by commercial transport from Bremen to Coblenz.

Telegraph.

According to the stipulations of the peace treaty, the Germans are not permitted to operate or install wireless apparatuses within the occupied zone. However, the American High Command, through the D.G.C.R.A., showed its willingness to authorise the use of very feeble and low power instruments for instructional purposes in the various provincial schools, to be subject to the inspection of the Signal Corps at such times as might be thought proper. A similar attitude was eventually adopted by the other armies.

Requests by the German authorities for permission to construct new telephone and telegraph lines continued throughout the year. A careful watch was kept on all such constructions, it being the opinion of several of the Allied authorities that many of those asked for might be considered almost entirely of a military nature.

Officers.

Major G. G. Bartlett, Infantry, served as American Delegate on the I.A.R.C., the I.A.L.C. and the D.G.C.R.A. from the beginning of the year to July 31st, 1922, his place on all three commissions being taken by

Major C. S. Caffery, Infantry. Major Caffery was relieved from duty on January 23rd, 1923.

Capt. W. H. Sadler, Q.M.C., continued to act as President of the Coblenz Delegation of the Railways Commission throughout 1922 and up to January 27th, 1923.

CHAPTER IV.

AMERICAN FORCES IN GERMANY.

Sub-Chapters

----- ooo O ooo -----

SUB-CHAPTER A.

COMPOSITION AND STRENGTH.

Command

The year 1922 brought no change in the command of the A.F.G. General Allen, who assumed command on July 8th, 1919, remained at the head of the American Forces until January 27th, 1923, when, following the President's decision to withdraw these forces, he handed over the command of the American area and the Coblenz bridgehead to General Marty, designated by General Degoutte to take over the old American area.

The promotion of Colonel W. H. Hay, G.S., to the rank of Brigadier General, and his subsequent departure for the U.S., vacated the post of Chief of Staff in 1922. This was temporarily filled by Lt. Col. H. H. Sheen from June 20, 1922 to July 31st, 1922, at which time Col. W. W. Harts, F.A., was announced as Chief of Staff. He served in that capacity until the end of the American occupation.

Organization.

The organization of the A.F.G. on January 1st, 1922, is shown in the following table:

O R G A N I Z A T I O N S

AMERICAN FORCES IN GERMANY

HEADQUARTERS

HEADQUARTERS A.F.G.

ENGINEERS
Co. "A", 16th Engineer Bn.
Co. "A" 1st Engineers

ENGINEERS
Co. "B" 1st Engineers
Co. "D" 1st Engineers

QUARTERMASTER
Supply Service
Bendorf
Coblenz
Coblenz-Lutzel
Andernach
Mayen
Port of Antwerp, Belgium
Detachment Transportation Service
Motor Transport Service

ARTILLERY
2ND Battalion, 6th Field Artillery.

ORDNANCE
18th Ordnance Co.
60th Ordnance Co.
82nd Ordnance Co.
82nd Ordnance Co. (Inactive)

MEDICAL
Station Hospital
Medical Supply Depot
Prov. Medical Bn.
Station Vet. Hosp.

AIR SERVICE
Detachment

SIGNAL CORPS
20th Service Co.

CAVALRY
Prov. Cav. Squadron.

HEADQUARTERS COMMANDANT.

Headquarters Detachment
Postal Detachment
Provost Marshal's Department and Provost Marshal
Prov. Military Police Co.

Disciplinary Barracks
Det. 8th Inf.
Military Prison
Det. 8th Inf.

1st BRIGADE, A.F.G.

HEADQUARTERS
Brigade Headquarters

INFANTRY
5th Infantry Regt.
8th Infantry Regt.

Station List January 1, 1922.

The stations of the various organizations on the same date may be seen from the following:

A M E R I C A N F O R C E S I N G E R M A N Y

HEADQUARTERS, American Forces in Germany	Coblenz
HEADQUARTERS Commandant	Coblenz
HEADQUARTERS, 1st BRIGADE, A.F.G.	Coblenz

CAVALRY	
Prov. Cavalry Squadron	Ehrenbreitstein
Headquarters Detachment	Ehrenbreitstein
Troops "A" & "B"	Ehrenbreitstein
INFANTRY	
Headquarters 5th Infantry	Andernach
Headquarters Company	Andernach
Service Company	Andernach
Howitzer Company	Andernach
1st Battalion (less Company "D")	Andernach
Company "D"	Engers
2nd Battalion (less Company "H")	Mayen
Company "H"	Engers
3rd Battalion (less Company "M")	Asterstein
Company "M"	Engers
Headquarters 8th Infantry	Coblenz
Headquarters Company	Coblenz
Service Company	Coblenz-Luetzel
Howitzer Company	Coblenz
1st Battalion (less Company "D")	Coblenz
Company "D"	Ehrenbreitstein
2nd Battalion (less Company "H")	Coblenz
Company "H"	Ehrenbreitstein
3rd Battalion (less Company "M")	Coblenz
Company "M"	Ehrenbreitstein
Medical Detachment	Coblenz
Disciplinary Barracks Det.	Coblenz
Military Prison Det.	Coblenz
Headquarters Detachment	Coblenz
Postal Detachment	Coblenz
Guard Detachment	Antwerp, Belgium
Provisional Military Police Co.	
Headquarters	Coblenz
Detachment	Andernach
Detachment	Wiesbaden
Detachment	Cologne
Detachment	Mayen
Detachment	Antwerp, Belgium
Detachment	Aix-la-Chapelle
Detachment	Bonn
Detachment	Boppard
Detachment	Bad Ems
Detachment	Niederlahnstein
Detachment	Montabaur
ARTILLERY	
2nd Battalion, 6th Field Artillery	Coblenz-Luetzel
Headquarters Detachment Combat Train	Coblenz-Luetzel
Batteries "D", "E" & "F"	Coblenz-Luetzel

ENGINEERS

Co. "A", 16th Engineer Battalion	
Headquarters & Supply Platoon	Coblenz
1st Platoon	Kreuzberg
2nd Platoon	Kottenheim
Co. "A" 1st Engineers	Coblenz
Co. "B" 1st Engineers	Coblenz
Co. "D" 1st Engineers	Coblenz

SIGNAL CORPS

20th Service Company	Coblenz

QUARTERMASTER

Supply Service	
Post Quartermaster	Coblenz
Depot "	Coblenz-Luetzel
Detachment	Bendorf
Detachment	Andernach
Detachment	Mayen
Port of Antwerp	Antwerp, Belgium

All stations in Germany unless otherwise noted.

Detachment Transportation Service.

Headquarters	Coblenz
Detachment	Hoenningen
Detachment	Remagen
Detachment	Elbingen
Detachment	Niedererbach
Detachment	Wiesbaden
Detachment	Cologne
Detachment	Mayen
Detachment	Andernach
Detachment	Engers
Detachment	Aix-la-Chapelle
Detachment	Wasserbillig
Detachment	Brussels, Belgium
Detachment	Antwerp, Belgium
Detachment	Bendorf
Detachment	Puderbach
Detachment	Paris, France.

Motor Transport Service

Headquarters	Coblenz-Luetzel
M.T.Co. No. 60	Coblenz
M.T.Co. No. 63	Coblenz
M. Rep. S. No. 100	Coblenz-Luetzel

MEDICAL

Station Hospital	Coblenz
Hospital Co. No. 60	Coblenz
Hospital Train No. 70	Coblenz
Medical Supply Depot	Coblenz
Station Veterinary Hospital	Coblenz-Luetzel.

Prov. Medical Battalion	Neuwied
Ambulance Co. No. 57	Neuwied
Hospital Co. No. 57	Neuwied
Hospital Co. No. 58	Neuwied

<u>ORDNANCE</u>

18th Ordnance Co.	Metternich
60th Ordnance Co.	Muelheim
82nd Ordnance Co. (inactive)	Metternich

<u>AIR SERVICE</u>

Detachment	Weissenthurm

<u>FINANCE</u>

Detachment	Coblenz

<u>Miscellaneous Units</u>

Civil Affairs	Coblenz
Civil Affairs	Mayen

P O S T O F F I C E S

A.P.O. No. 927	Coblenz
A.P.O. No. 944	Antwerp, Belgium

R E C R U I T I N G S T A T I O N S

Germany - Station Hospital	Coblenz
Belgium - Hdqrs., Port of Antwerp	Antwerp, Belgium

All stations in Germany unless otherwise noted.

<u>Organization and Station List After July 1st, 1922.</u>

The radical reduction in the strength of the A.F.G. ordered and affected during the first half of the year necessitated many shifts, combinations and changes in the organization of the forces. Some units were returned to the States intact, thus being dropped from the organization altogether. The changes may best be seen by a comparison between the organization as shown on January 1st, 1922, and the following table, giving the organization adopted as soon as practical after the reductions ceased, and which was continued without material change until all the troops were withdrawn:

Organizations	Station
Headquarters	Coblenz
Quartermaster	Coblenz
Q.M. Det.	Coblenz
M.R.S. # 100	Coblenz
Station Hospital	Coblenz
Co. "A" 16th Engr. Bn.	Coblenz
20th Service Co., S.C.	Coblenz
8th Inf. (less 1 bn.)	Coblenz

French Troops.

The most important change in the organization of the A.F.G. occurred in April, 1922, when, with the consent of the Commanding General, certain French troops arrived in the American zone for station as a part of General Allen's command. The effort to include a French contingent in the American zone dates back to the days immediately following the armistice, when a request to that effect from Marshal Foch was, with the sanction of the President, refused by General Pershing. The division of the original Coblenz bridgehead as a result of this refusal, the events leading to General Allen's decision to allow this move in the spring of 1922, when it was thought that all the Americans would be withdrawn, and the many unsuccessful efforts on the part of the French to increase and enlarge their holdings in the American zone after they finally got in, would alone make an interesting chapter.

The first French troops, consisting of a small cavalry detachment, arrived in the area on April 19th, and the French garrison was properly established by July. The following table shows the French strength by months:

1922	Officers	Enlisted Men.
July 31	122	4449
August 30	127	4537
September 30	130	4240

October 31	161	4680
November 30	173	5425
December 31	174	5399
1923		
January 27	174	5399

The following orders are quoted as having had an important effect on the organization of the American Forces in Germany during 1922:

"GENERAL ORDERS NO. 35 12 May, 1922,

COMMAND IN THE COBLENZ AREA.

The following approved order of the French Army on the Rhine is published for the information of all concerned:

I. All French troops stationed in the American zone are placed under the supreme command of General Allen, Commanding General, American Forces in Germany, who will decide as to the employment of these troops and exercise the district command.

II. On and after May 9, the Commanding General, 4th Brigade of Dragoons, will fulfil the duties of garrison commander as regards the French troops, whether stationed at Coblenz or at neighboring stations. In this connection he will have such rights and powers over them as are vested in garrison commanders by the Garrison Service Regulations, with the proviso above all that he shall comply with the orders and instructions received from General Allen in his capacity as district commander, especially as regards orders issued by the American authorities on the subject of prostitutes, sale of alcohol and police regulations.

III. As garrison commander of the French forces in the American zone, the Commanding General of the 4th Brigade of Dragoons will be subordinate to:

1. General Allen as regards:
    a) carrying out of provisions issued for the district by that general officer and reports and statements in connection therewith:
    b) taking charge of barracks given to the French Army;
    c) policing of French troops stationed in the American zone.

The Commanding General of the 4th Brigade of Dragoons will have the services of Captain Hendrickx,

French Liaison Officer, at his disposal, in order to assist him in his relations with the Americans authorities.

2. The Commanding General of the 30th Army Corps as regards:

a) general discipline and mobilization of the 156th infantry Regiment.
b) military jurisdiction over all French troops stationed in the American zone. The American courts-martial at present existing, however, will of course continue to exercise jurisdiction over the civilian population in all matters regarding relations between the French troops stationed in the American zone and the population.

3. The Commanding General of the 4th Cavalry Division in all questions in connection with the Brigade of Dragoons and any others excepting those under General Allen.

IV. The powers of the French garrison commander of Coblenz in the matter of billets and barracks will be dealt with in subsequent instructions.

DEGOUTTE,
Commanding General
Army of the Rhine.

By command of Major General ALLEN:

H. H. SHEEN,
Acting Chief of Staff".

"GENERAL ORDERS NO. 55. 25 July, 1922.

COMMANDING OFFICER, FRENCH TROOPS, AMERICAN AREA.

1. The Senior Line Officer of the Army of Occupation, French Army of the Rhine, on duty in the American area, will fulfil the duties of Garrison Commander of all French troops stationed in the American area. The designation of this officer will be 'Commanding Officer, French Troops, American Area'.

2. The Commanding Officer, French Troops, American Area, will report to these Headquarters (Attention A.C. of S., G-1) three days in advance, the date and hour of expected arrivals of any troops of the French Army. On the morning following the arrival of any such troops, he will report to these Headquarters the number of officers, men, animals, guns and vehicles, with each organization, and the names of all officers present herewith. Before the departure of any organization or detachment, he will report the proposed date and hour of departure, and will make an inspection of sanitary conditions

of barracks, stables and grounds to be vacated. After the departure, he will report the number of officers, men, animals, guns and vehicles departing, with the names of all officers who have left the area with the detachment.

3. The Commanding Officer, French Troops, American Area, will submit a weekly report to the Adjutant, American Forces in Germany, so as to reach these Headquarters not later than 10:00 A.M., Monday of each week, showing the number of officers, men, animals, guns and vehicles with each organization on duty in the American Area as of midnight the preceding Sunday.

4. The Commanding Officer, French Troops, American Area, will be held responsible that all organizations coming into the American area are furnished with existing orders of the American Forces in Germany which apply to the French troops stationed in this area. He will take measures to insure compliance with such orders by all French troops.

By command of Major General ALLEN:

H. H. SHEEN,
Acting Chief of Staff."

Strength.

In accordance with the approved plan for the reorganization of the Army at reduced strength, and pursuant to instructions received from the War Department late in 1921 directing the placing of one Infantry regiment of the A.F.G. on the inactive list, the demobilization of certain provisional organizations and the reduction of the remaining units to a specified strength, these forces had undergone by January 1st, 1922, such changes that the assigned and authorized personnel of the command on that date stood as follows:

ASSIGNED AND AUTHORIZED STRENGTH
AMERICAN FORCES IN GERMANY AS OF 1 JAN. 1922.

	Officers		Enlisted men.	
	Assgd.	Authrd.	Assgd.	Authrd.
Headquarters, AFG	84	74		
Quartermaster		23		500
Supply Service	14		495	
Motor Transport Service	14		139	
Det. Trans. Service	9		107	
Port of Antwerp	12	9	64	

ASSIGNED AND AUTHORIZED STRENGTH

AMERICAN FORCES IN GERMANY AS OF 1 JAN. 1922.
(continued)

	Officers		Enlisted men	
	Assgd.	Authrd.	Assgd.	Authrd.
Medical Department		61		400
Station Hospital	36		435	
Medical Supply Depot	1		11	
Sta. Vet. Hospital	0	0	13	
Finance Department	0	(5)	15	22
Prov. Cavalry Squadron	10	8	226	131
2nd Bn. 6th F.A.	19	21	515	440
Co. A. 16th Engr. Bn.	5	3	134	90
Co. A. 1st Engrs.	3	3	135	85
Co. B. 1st Engrs.	3	3	113	85
Co. D. 1st Engrs.	4	3	119	85
18th Ordnance Company	2	2	70	48
20th Service Co. (S. C.)	4	3	202	150
60th Ordnance Company	3	3	109	75
Air Service Detachment	10	12	131	88
Total	233	233	3033	2199

By Units

Headquarters Commandant				
Hqrs. Commandant, Hdqrs.	8	2		
Headquarters Detachment	5	4	470	300
Provost Marshal &P.M.D.	3	1		
Prov. Military Pol. Co.	5	4	298	160
Mil. Pris. Det., 8th Inf.	2	1	11	
Postal Det., Hq. Det.	1	1	0	0
Total	24	13	779	460
1st Brigade, A.F.G.				
Brigade Headquarters	7	6	62	34
5th Infantry	79	59	1762	1262
8th Infantry	73	61	1645	1262
Total	159	144	3469	2558
Grand Total	416	372	7281	5217

Reduction

The year 1922 was destined to be one of reduction as far as the Rhine forces were concerned. On February 1st the War Department not only disapproved certain requests for additional personnel and ratings but directed a further reduction, this time in the enlisted strength of the Quartermaster Corps. On February 27th another cut of 203 officers and 3,000 men was ordered, which left the A.F.G. with only one infantry regiment, one battery of artillery and no air service. Other branches and arms were so reduced as to make proper functioning of a composite force a matter of great difficulty. Later, in a War Department letter dated February 18th, the above figures were changed to read 210 officers and 2,861 men, which left the authorized strength of the A.F.G. at 162 officers and 1,926 men.

Following close upon these orders came a cable from the War Department stating that the President had decided to withdraw all the American troops from the Rhine and directing that the entire movement be completed by July 1st. Various political factors both at home and abroad brought about a change in the administration's plan, however, and on June 4th, the Commanding General was authorised to retain a composite force not to exceed 1,200 officers and men on the Rhine after June 30th. This authorization was received after all the artillery units and one battalion of the remaining infantry regiment (8th Inf.) had departed for the United States. In order that the various orders of reduction may be readily seen, a condensed table is given below:

a-- officers.
b-- enlisted men.

Column 1 - Letter October 15, 1921.
2 - Letter February 1, 1922.
3 - 1st Cable sent - W-883.
4 - 2nd Cable sent W-886 -- Letter February 18, 1922.
5 - Resulting Authority May 1, 1922.
6 - July 1, 1922, to January, 1923.

Unit	1		2		3		4		5		6	
	a	b	a	b	a	b	a	b	a	b	a	b
5th Inf.	55	1312	55	1312	55	1312	55	1312	-	-	-	-
8th Inf.	55	1312	55	1312	-	-			55	1312	46	885
Brig. Hq.	6	34	6	34	6	34	6	34	-	-		
Bn. 6th F.A.	20	452	20	452	16	349	16	349	4	103		
Cos. A.B.D.- 1st Engrs.	9	270	9	270	6	180	6	180	3	90	-	-
Det. A.S.	13	88	13	88	13	88	13	88	-	-	-	-
Sig. Co.	6	150	6	150	3	75	3	75	3	75	1	35
M. Rep. Sec. 100	-	-	-	-	-	-	-	-	-	-	2	18
Q. M. C.	27	500	27	354	7	300	14	250	13	104	9	38
M. D.	60	400	60	400	38	318	38	318	22	82	8	29
Ord. Dept.	5	123	5	123	1	75	1	75	4	48	1	3
Fin. Dept.	5	22	5	22	-	-	-	-	5	22	2	4
Hq. Tr.	3	69	3		3	69	3					
Hq. A.F.G.	105	395	105	-	55	200	55	-	50	-	32	
Ry. Engr. Bn.	3	90	3	90	-	-	-	-	3	90	3	75
Total	372	5217	372	4607	203	3000	210	2681	162	1926	104	1087

Various other Washington cables and letters directing reorganization of certain units, redistribution of grades and ratings, re-allocation of personnel by grades and specialist ratings in various branches and arms, etc., are not included in this narrative, as compliance therewith did not materially change the strength after June 30th.

The next important cable from the War Department bearing on strength was not one of reduction but one of withdrawal. On January 11th, 1923, instructions were received directing the withdrawal of the entire American Forces in Germany. Steps necessary to effect this movement were taken at once and on January 25th, the U.S.A.T. "St. Mihiel", sent from the United States for the purpose, sailed from Antwerp, returning all of the combatant troops of these forces to the United States.

The following table shows the progressive reduction of the A.F.G. and gives a clear picture of the strength situation.

This table shows the strength of the A.F.G. at the end of each month, from January, 1922, to January, 1923, both inclusive:

1922	Officers	W.O.	Nurses	A.F.G.	Enl. Men	Animals
January	417	20	42	28	7294	1922
February	415	20	42	28	7265	1937
March	400	20	41	28	6185	1939
April	326	18	39	26	4240	1239
May	249	17	17	24	2634	782
June	249	17	17	24	2634	782
July	113	10	2	14	1103	514
August	113	10	4	14	1103	517
September	113	10	4	14	1091	469
October	114	10	4	14	1095	468
November	113	10	4	13	1047	466
December	107	10	4	13	1058	465
Jan. 1, 1923	106	10	4	13	1060	465
Jan. 31, 1923	58	8	4	10	191	445

To supplement the above, it might be added that on January 1st, 1923, the enlisted strength of the A.F.G. was classified as follows:

Q.M.C.	Med. Dept.	Sig. Corps	Eng.	F. D.	Inf.	Ord.	Cav.
66	30	33	78	4	845	3	0

It is interesting to note that during the calendar year 1922, these forces were reduced by 267 officers and 7,162 enlisted men; that the largest number of army passengers, including officers, men, women and children, was evacuated on the U.S.A.T. "Cantigny" in February, 1922, and that during the month of May - from the 16th to the 27th, a period of 12 days - four transports engaged in the evacuation, returning to the United States 92 officers (exclusive of dependents), 1,050 enlisted men and 300 women and children (soldiers' dependents, 2nd class).

This reduction was not without its selective element, and whenever possible and consistent with preservation of the unity of the organizations, the officers and men to be retained on the Rhine were selected for their special qualifications for service with these forces.

Officers - Adjutant's Office.

Pursuant to the general policy of reduction, the office of the Adjutant lost two of its commissioned personnel. Major C. K. Nulsen, Assistant Adjutant, was relieved on June 7th, 1922, and assigned to the 8th Infantry. Major C. M. Thiele, Assistant Adjutant, was relieved on June 12th, 1922, and assigned to the Office of Civil Affairs. The following officers remained on duty in the office of the Adjutant until the withdrawal of the American forces:

Major E. G. Talbott, A.G.D., Adjutant.
Captain W. E. Lauer, Inf., Assistant Adjutant.

French Field Officers.

The following list shows the field officers on duty with French troops under the jurisdiction of the Commanding General, A.F.G., at the time that the American area was turned over to the French command:

Majors:	Bon, Etienne	230th F.A.
	Cavard, Eugene J.	156th Inf.
	Chanal, Jean M.	151st Inf.
	Couillebault, Olivier	Fin. Dept.
	Dardelet, Hugues	151st Inf.
Lt. Col.	De Bertuche De Mendite, M. C.	156th Inf.
Majors:	Delpit, Jean H.	20th F.A.
	De Moras, Marius	151st Inf.
	Dreyfres, Achille	M.C.
	Dryjard, des Garniers, Jacques	156th Inf.
Lt. Col.	Du Roselle, Charles	M.C.
Major	Guyot, Siounest	230th F.A.
Lt. Col.	Dumont, Victor	230th F.A.
Major	Islert, Jules	156th Inf.
Colonel	Kiffer, Jean	
Major	Knecht, Jean B.	156th Inf.

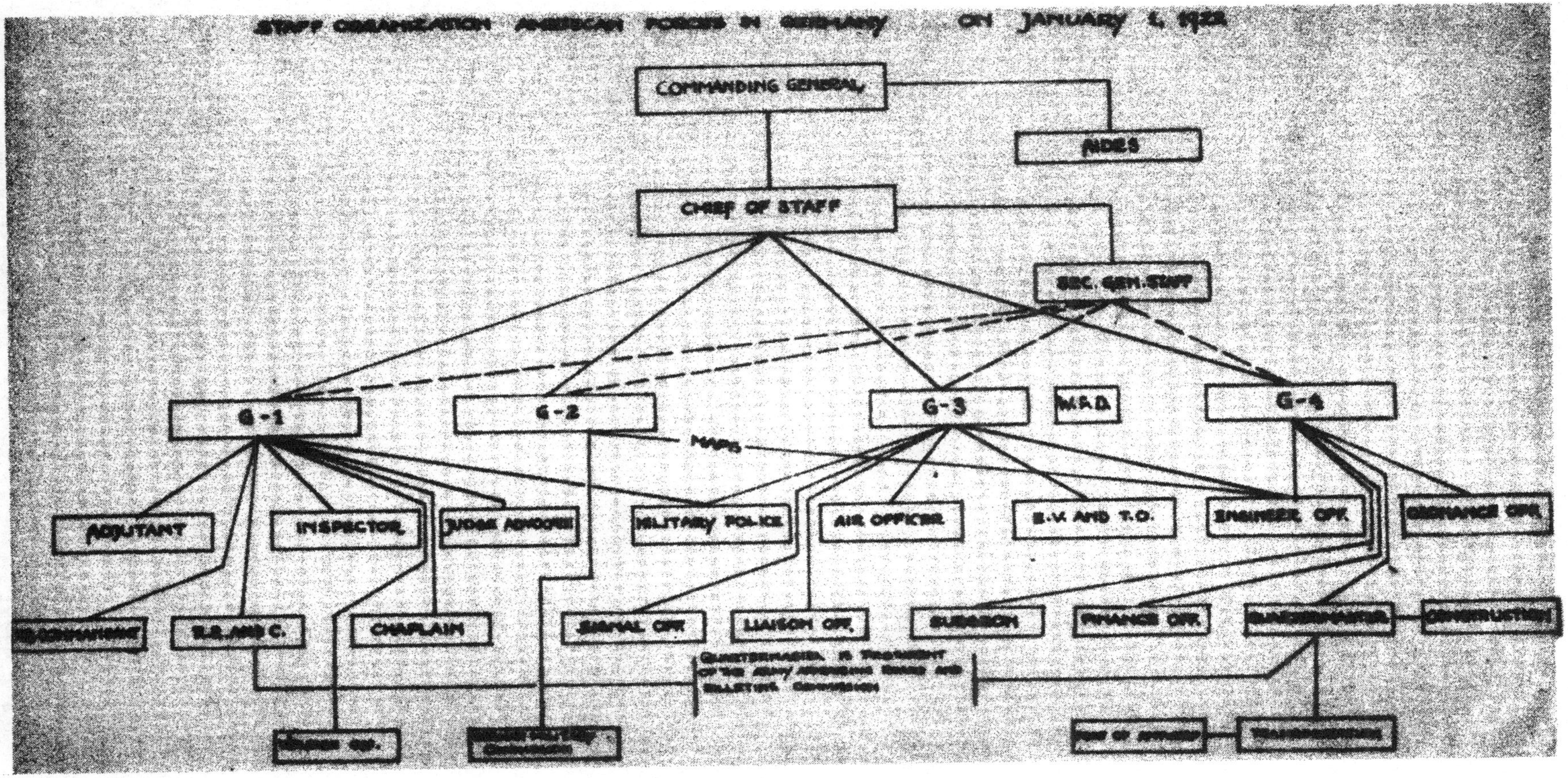
STAFF ORGANIZATION AMERICAN FORCES IN GERMANY
COMMANDING GENERAL
AIDES
CHIEF OF STAFF
SEC. GEN. STAFF
G-1
G-2
G-3
G-4
MAPS
ADJUTANT
INSPECTOR
JUDGE ADVOCATE
MILITARY POLICE
AIR OFFICER
ENGINEER OFF.
ORDNANCE OFF.
CHAPLAIN
SIGNAL OFF.
LIAISON OFF.
SURGEON
FINANCE OFF.
QUARTERMASTER
CONSTRUCTION
TRANSPORTATION

Major	Koblick, Georges J.	156th Inf.
Lt. Col.	Le Merre, Louis	156th Inf.
Maj.	Luquet, Armand	151st Inf. M.C.
Brig.Gen.	Marty, Raoult, M.	154th Brigade
Maj.	Mayot, Anne	156th Inf.
Maj.	Mesnil, Louis	151th Inf.
Maj.	Pellet, Charles J.	Med. Det.
Col.	Piazza, Pierre	156th Inf.
Lt. Col.	Raoult, Louis	151st Inf.
Maj.	Roussotte, Felix	151st Inf.

DROPPED
Between July, 1922, and January, 1923.

Maj. Bounefoces	30th Dragoons
Lt. Col. Brice	30th Dragoons
Brig. Gen. de Partouneaux	4th Dragoons
Maj. de Laundraus R.	151st Inf.
Lt. Col. Gegonne, S.	151st Inf.
Lt. Col. Goursault	28th Dragoons
Col. Kirschleger	28th Dragoons
Maj. Mcinot-Verly	28th Dragoons
Col. Vauniere, Paul	151st Inf.

ALPHABETICAL ROSTER OF OFFICERS ON DUTY
WITH THE A.F.G. FROM
1st JANUARY, 1922 to 3rd FEBRUARY, 1923.

NAME	RANK	ORGANIZATION
Allen, Henry T.	Maj. Gen. USA	Commanding General 1st Jan. '22 to date.
Allen, Henry T. Jr.	Capt. Cav.	Cav. Sq. & Hq. AFG 1st Jan. '22 to date.
Allen, Virgil G.	1st Lt. Inf.	Co. F. 5th Inf, 1st Jan. '22 to 7 Apr. '22,
Almquist, Elmer H.	Capt. F.A.	2nd Bn. 6th F.A. 1st Jan. '22 to 22 Apr. '22.
Andrews, Frank M.	Maj. A.S.	Air Officer & Civ. Aff. 1st Jan. '22 to date.
Ankcorn, Charles M.	Capt. Inf.	8th Infantry 1st Jan. '22 to 25 Jan. '23.
Archer, Simon D.	1st Lt. QMC	Det. Trans. Serv. 1st Jan. '22 to 16 May '22.
Armstrong, Walter C.	Capt. Inf.	Serv. Co. 8th Inf. 1st Jan. '22 to 18 Mar. '22.
Arnold, James W.	Capt. Inf.	8th Infantry. 1st Jan. '22 to 25 Oct. '22.
Arthur, Dogan H.	1st Lt. AS	Air Serv. Det. 1st Jan. '22 to 22 Apr. '22.
Austin, Francis M.	1st Lt. V.C.	Sta. Hosp. Coblenz. 3d Feb. '22 to 4 Dec. '22.
Avery, Clark M.	1st Lt. Inf.	Hq. 8th Infantry 1st Jan. '22 to 10 April. '22.
Avery, Ray L.	Maj. Ord.	18th Ord. Co. 1st Jan. '22 to 26 May '22.
Babbitt, George M.	Capt. DC	2nd Bn. 5th Inf. 1st Jan. '22 to 18 Mar. '22.
Babbitt, Richard C.	1st Lt. Inf.	Co. A. 5th Inf. 1st Jan. '22 to 9 Mar. '22.
Backman, Stanley G.	Capt. Inf.	Co. L. 5th Inf. 1st Jan. '22, to 7 Apr, '22.
Bagby, Philip H.	Maj. GS (Inf)	A.C. of S. G-2, AFG 1st Jan. '22 to 16 Sept. '22.
Bailey, Cornelius O.	Maj. M.C.	Hq. Comdt. & Hq. Det. 1st Jan. '22 to 30 June '22.
Baker, Arthur A.	Capt. Inf.	8th Infantry 1st Jan. '22 to 4 Oct. '22.
Baker, Joseph K.	1st Lt. Cav.	Cav. Sq. & Hqrs. AFG 1st Jan. '22 to 25 Jan. '23.
Barnes, Gladeon M.	Maj. Ord.	Off. Ord. Off. Hq. AFG 26th Apr. '22, to 29 Jan. '23.
Barroll, Morris K. Jr.	Capt. QMC	RTO Coblenz Det. T. Serv. 1st Jan. '22 to date.
Barth, George B.	1st Lt. Inf.	8th Infantry. 1st Jan. '22 to 25 Jan. '23.

Bartlett, George G.	Maj. Inf.	Lia. Off. Wiesbaden	1st Jan. '22 to 31 July '22.
Barton, Raymond O.	Maj. Inf.	8th Infantry.	1st Jan. '22 to 25 Jan. '23.
Bartow, Donald M.	Capt. Inf.	Co. A. 5th Inf.	1st Jan. '22 to 9 Mar. '23.
Bates, Walter T.	Col. Inf.	Comdg. 8th Infantry.	1st Jan. '22 to 25 Jan. '23.
Bauman, Joseph W.	Maj. M.C.	Station Hosp. Coblenz.	18th Feb. '22 to 16 Apr. '22.
Beard, Paul B.	Capt. QMC	Motor Trans. Serv.	1st Jan. '22 to 30 June, '22.
Bell, Robert P.	Capt. Inf.	Postal Officer, AFG	1st Jan. '22 to 20 June, '22.
Bell, Virgil	Capt. Inf.	8th Infantry.	1st Jan. '22 to 25 Jan. '23.
Bender, Walter	Capt. A.S.	Air Serv. Det.	1st Jan. '22 to 22 Apr. '22.
Bennett, Walter S.	Capt. M.C.	2nd Bn. 6th F.A.	1st Jan. '22 to 22 Apr. '22.
Bethel, Edwin A.	Maj. Engrs.	Engr. Off. Hq. AFG	1st Jan. '22 to 10 Dec. '22.
Bevans, Stuart M.	1st Lt. FA	2nd Bn. 6th F.A.	1st Jan. '22 to 22 Apr. '22.
Beyette, Hubert W.	Capt. QMC.	M.R.S. #100	28th Nov. '22 to date.
Biggar, John	Capt. QMC	Pt. Antwerp & Hq AFG.	1st Jan. '22 to date.
Blanchard, Charles C.	1st Lt. F.A.	2nd Bn. 6th F.A.	1st Jan. '22 to 22 Apr. 1922.
Blaney, John	1st Lt. AS	Air Serv. Det.	14th Feb. '22 to 22 Apr. '22.
Bledsoe, William P.	Capt. F.A.	2nd Bn. 6th F.A.	1st Jan. '22 to 22 Apr. '22.
Bodeker, Charles C.	Capt. Inf.	PR&C Off. Hq. AFG.	1st Jan. '22 to 26 May '22.
Bollenbeck, Joseph W.	Capt. Inf.	2nd Bn.	1st Jan, '22 to 2 Feb. '22.
Bond, Easom J.	1st Lt. Inf.	8th Inf.	1st Jun. '22 to 25 Jan. '23.
Breckinridge, C.R. Jr.	Capt. Inf.	Prov. M.P. Co.	1st Jan. '22 to 30 June '22.
		Town Major Coblenz.	14th Sept. '22 to date.
Brennan, Francis M.	Capt. Inf.	8th Infantry,	1st Jan. '22 to 25th Jan. '23.
Brinkley, Thomas M.	1st Lt. Inf.	Co. C. 5th Inf.	1st Jan. '22 to 9 Mar. '22.
Brockway, Edgar K.	Capt. Inf.	Co. K. 5th Inf.	1st Jan. '22 to 7 Apr. '22.
Broedlow, Rudolph W.	1st Lt. Inf.	Serv. Co. 5th Inf.	1st Jan, '22 to 18 Mar. '22.
Brown, Frank W.	Capt. S.C.	20th Serv. Co. SC	1st Jan, '22 to 30 June '22.
		Actg. S.O. Chief Sig. O.	31st Aug. '22 to date.

Brown, Harold O.	Capt. M.C.	Prov. Med. Bn. 1st Jan. '22 to 18 May '22.
Browne, Ivins LaR.	1st Lt. Inf.	Co. K. 8th Inf. 1st Jan. '22 to 15 Nov. '22.
Browning, Robert M.	Capt. Inf.	Det. Tran. Serv. & 8 Inf. 1st Jan. '22 to 21st Apr. '22.
Buckley, Wm. R.	Capt. QMC	Off. Q.M. AFG 21st Nov. '22 to date.
Bunting, Loyd D.	1st Lt. Inf.	5th Infantry 1st Jan. '22 to 7 Apr. '22.
Burney, Joel R.	Capt. Inf.	Co. I. 5th Inf. 1st Jan. '22 to 19 Mar. '22. (AFG)
Burwell, Harvey B.S.	Maj. A.S.	Comdg. A.S. Det & Htt Hq 1st Jan. '22 to 20 June, '22.
Byrne, Francis A.	Capt. Inf.	Mil. Pris. Det. 1st Jan. '22 to 22 May '22.
Caffery, Charles S.	Maj. Inf.	Lia. O. Wiesbdn. Hq. AFG 31st July '22 to date.
Calladine, Thomas M.	Capt. M.C.	Sta. Hospital, Coblenz. 1st Jan. '22 to 16 May '22.
Campbell, Thomas E.	Capt. Inf.	Co. C., 5th Inf. 1st Jan. '22 to 9 Mar. '22.
Campbell, William F.	Capt. QMC.	M.R.S. #100 1st Jan. '22 to 7 Apr. '22.
Campbell, William R.	Capt. M.C.	Prov. Med. Bn. 1st Jan. '22 to 7 Apr. '22.
Cannon, Mimucan D.	Capt. Inf.	Co. M. 5th Inf. 1st Jan. '22 to 31 Jan. '22.
Carroll, James V.	1st Lt. FA.	2nd Bn. 6th F.A. 1st Jan. '22 to 22 Apr. '22.
Carroll, William J.	Capt. M.C.	Sta. Hosp. Coblenz. 1st Jan. '22 to 7 Apr. '22.
Case, Robert A.	Capt. Inf.	Co. D. 5th Inf. 1st Jan. '22 to 9 Mar. '22.
Case, Hugh J.	Capt. Engrs.	Co. A 1st Engrs & Off Engr Off. Hqrs. 1st Jan. '22 to 13 May '22.
Cauthorn, Francis S.B.	Capt. Inf.	8th Infantry 1st Jan. '22 to 25 Jan. '23.
Cawthorne, Ted H.	Capt. Inf.	8th Infantry. 1st Jan. '22 to 16 May. '22.
Chapman, Gustavus F.	1st Lt. QMC	QM Depot Cobl-Lutz. 1st Jan. '22 to 30 June '22. Off of the QM. AFG 31st Aug. '22 to date.
Chester, Martin S.	Capt. Inf.	Off. Hq. Comd't. AFG 1st Jan. '22 to 18 Mar. '22.
Chickering, William E.	Capt. Inf.	8th Infantry. 1st Jan. '22 to 25 Jan. '25.
Claussen, Walter C.	Capt. Inf.	Asst. Adj. 5th Inf. 1st Jan. '22 to 13 Mar. '22.
Cleary, Wm. D.	1st Lt. Chap.	Off. Chap. AFG 1st Jan. '22 to 10 Apr. '22. Off. Chap. AFG. 24 Jan. '22 to date.

Cleave, John W.	1st Lt. MAC	Off. Ex Off Surg AFG	1st Jan. '22 to 20 June '22.
Clemenson, Wendell L.	Capt. Inf.	8th Infantry.	1st Jan. '22 to 25 Jan. '23.
Cobb, Elton T.	1st Lt. Inf.	8th Infantry.	1st Jan. '22 18 Mar. '22.
Cobb, Murray A.	Capt. Inf.(NA)	Phy Recons. Switz.	1st Jan. '22 to 27 Apr. '22.
Coffin, Frank P.	Capt. Inf.	Co. E. 5th Inf.	1st Jan. '22 to 19 Mar. '22.
Coleman, LeVert	Col. CAC.	Lia. Off. Mainz	1st Jan. '22 to 31 Dec. '22.
Collins, Lawrence C.	1st Lt. Inf.	Co. M. 5th Inf.	1st Jan. '22 to 9 Mar. '22,
Collins, Patrick	1st Lt. Inf.	Co. G. 5th Inf.	1st Jan. '22 to 9 Mar. '22.
Conrad, George B.	1st Lt. Inf.	Co. G. 5th Inf.	1st Jan. '22 to 9 Mar. '22.
Cook, Gilbert R.	Maj. Inf.	8th Infantry.	1st Jan. '22 to 25 Jan. '23.
Cooper, James C. Jr.	1st Lt. Inf.	Det. Trans. Serv.	1st Jan. '22 to 20 June '22.
Corbin, Clifford L.	Maj. QMC	Post QM Coblenz.	1st Jan. '22 to 31 Aug. '22.
Cound, Oliver E.	1st Lt. QMC	QM Det. Bendorf	1st Jan. '22 to 18 May '22.
Crawford, Harry	1st Lt. F.A.	2nd Bn. 5th F.A.	1st Jan. '22 to 22 Apr. '22.
Crea, Harry B.	Maj. Inf.	Civil Affairs, Mayen & Coblenz.	1st Jan. '22 to 20 June '22.
Creed, Richard L.	Capt. Cav.	Prov. Cav. Sq & RR&C O.	1st Jan. '22 to date.
Crim, Casper R.	Capt. Inf.	5th Infantry	1st Jan. '22 to 7 Apr. '22.
Crump, Ira A.	Capt. O.D.(CAC)	60th Ord. Co. & Ord. O. AFG	1st Jan. '22 to date.
Dahlquist, John E.	Capt. Inf.	Co. G. 5th Inf.	1st Jan. '22 to 9 Mar. '22.
Dalton, Joe N.	Capt. Inf.	Adj. 8th Inf.	1st Jan. '22 to 10 May '22.
		8th Infantry,	4th Dec. '22 to 2 Jan. '23.
		A.D.C. & Sec. G.S.	2nd Jan. '23 to date.
Dashiell, Leland R.	Capt. M.C.	Surg. 2nd Bn. 8th Inf.	1st Jan. '22 to 18 May '22.
Davis, Arthur O.	Maj. M.C.	Sta. Hosp. Coblenz.	1st Jan. '22 -16 May '22.
Davis, Orville E.	1st Lt. QMC	Off Q.M.AFG	21 Nov. '22 to 25 Jan. '23.
Davis, Thomas J.	1st Lt. Inf.	D.C. I. Officer.	1st Jan. '22 to date.
Dawson, Allen W.	Capt. M.C.	Surg. 5th Inf. Andernach	1st Jan. '22 to 18 Mar. '22.
Dear, William R.	Maj. M.C.	Station Hospital, Cobl.	1st Jan. '22 to 20 Feb. '22.

Deck, Frederick W.	Capt. Engrs.	Co. B. 1st Engrs.	1st Jan. '22 to 18 Mar. '22.
Dedicke, Alfred E.	Capt. Inf.	G-2 Sec. Hq. AFG	1st Jan. '22 to 25 Jan. '23.
Denholm, Charles J.	Capt. D.C.	Off Att Surg Sta. Hosp.	1st Jan. '22 to 18 May '22.
Dever, Wesley O.	1st Lt. F.A.	2nd Bn. 6th F.A.	1st Jan. '22 to 22 Apr. '22.
Devine, Mark A. Jr.	1st Lt. Cav.	Prov. Cav. Sq & Hq AFG	1st Jan. '22 to date.
Dibble, John	Maj. M.C.	Surg. Air Serv. Det.	1st Jan. '22 to 18 May '22.
Dietz, Lawrence A.	1st Lt. Inf.	8th Infantry	1st Jan. '22 to 25 Jan. '23.
Distelhorst, Fred M.	Capt. Inf.	Hqrs. Det. AFG	1st Jan. '22 to 30 June '22.
Doane, Guy R.	Maj. F. D.	Fin. Off. Antwerp Belg.	1st Jan. '22 to 20 June, '22.
Doll, Frederick S.	Capt. Inf.	8th Inf. & Hq. AFG	1st Jan. '22 to date.
Donaldson, Thomas Q. Jr.	1st Lt. Cav.	Prov Cav Sq & Hq Det. AFG	1st Jan. '22 to date.
Donaldson, Wm. H.	Capt. CAC	R R & C Serv. AFG	1st Jan. '22 to 10 Apr. '22.
Donoghue, Walter J.	1st Lt. Chap.	Chap. Hq. 5th Inf.	1st Jan. '22 to 9 Mar. '22.
Doran, Wm. T.	1st Lt. QMC	Det. Tran. Serv.	11 Feb. '22 to 25 May '22.
Downer, John W.	Maj. F.A.	G-3 Sec. AFG	1st Jan. '22 to date.
Downs, Sylvester D. Jr.	Maj. F.A.	Civ. Affairs, Coblenz.	1st Jan. '22 to 22 Apr. '22.
Drinkwater, John G.	Capt. Engrs.	Off. Engr. Off. Hq AFG	1st Jan. '22 to 7 Apr. '22.
Dunlop, Samuel L.	Capt. Inf.	5th Inf.	1st Jan. '22 to 7 Apr. '22.
Dunn, William McK.	Capt. F.A.	ADC & G-5 Sec. Hq AFG	1st Jan. '22 to 13 Nov. '22.
Early, John M.	1st Lt. A.S.	Air Serv. Det.	1st Jan. '22 to 18 May '22.
Easterbrook, Edmund P.	Chap. Lt. Col.	Senior Chap. AFG	1st Jan. '22 to date.
Ekman, George A.	2nd Lt. MAC	Prov Med. Bn. Sta Hosp.	1st Jan. '22 to 7 Apr. '22.
Elliott, Walter A.	1st Lt. Inf.	Hq Co. 5th Inf.	1st Jan. '22 to 9 Mar. '22.
Elliott, Wilbur B.	Capt. QMC	Det. Trans. Serv.	21st Feb '22 to 26 May '22.
Ellis, Edmund deT	Maj. QMC	QM Depot Coblenz-Lutz and QM Det. Coblenz.	1st Jan. '22 to date.
Elmendorf, DuMont F.	Capt. M.C.	5th Inf. & Sta. Hosp.	1st Jan. '22 to 9 Apr. '22.
Feamster, Claude N.	Maj. Inf.	RTO Andernach	1st Jan. '22 to 18 Mar. '22.

Fechet, D'Alary	Maj. Inf.	Ex. O. Hq. 1st Brig.	1st Jan. '22 to 18 May '22.
Feldman, Herman	Capt. F.A.	2nd Bn. 6th F.A.	1st Jan. '22 to 22 May '22.
Fellows, Harold C.	Capt. Cav.	Hqrs. Comd't Off.	1st Jan. '22 to 26 May '22.
Finley, David H.	Capt. Inf.	Prov. M.P. Co. (H.C.)	1st Jan. '22 to 18 May '22.
Fitts, Francis M.	Maj. M.C.	Sta. Hosp. Coblenz.	1st Jan. '22 to 18 May '22.
Flynn, Thomas J.	Maj. M.C.	Civil Affairs, Cobl.	1st Jan. '22 to 26 May '22.
Fortier, Malcolm V.	Capt. Inf.	8th Infantry.	1st Jan. '22 to 25 Jan. '23.
Foster, Robert J.	Maj. V.C.	Vet. Officer AFG	1st Jan. '22 to 20 June '22.
Foster, Robert T.	1st Lt. Inf.	8th Infantry.	1st Jan. '22 to 25 Jan. '23.
Francis, Frank D.	Maj. M.C.	Station Hosp. Coblenz.	9 Mar. '22 to date.
Gantt, Henry P.	Capt. F.A.	2nd Bn. 6th F.A.	1st Jan. '22 to 22 Apr. '22.
Gardner, Harry F.	Capt. QMC	QM Depot Cobl-Lutz.	1st Jan. '22 to 30 June '22.
Gayle, Richard F.	Capt. Inf.	2nd Bn. 8th Inf.	1st Jan. '22 to 16 May '22.
Gerow, Lee S.	Capt. Inf.	G-1 Section, AFG	1st Jan. '22 to date.
Gilbert, Clarence L.	Maj. CAC	Off. Q.M. AFG	1st Jan. '22 to 1 Dec. '22.
Gill, William D.	Capt. M.C.	Station Hosp. Cobl.	1st Jan. '23 to 26 May '22.
Gillespie, Francis J.	1st Lt. Inf.	8th Infantry.	1st Jan. '22 to 25 Jan. '23.
Glasgow, Joseph H.	1st Lt. Inf.	Hqrs. and ADC CG AFG	1st Jan. '22 to date.
Goodrich, Paul M.	Lt. Col. Inf.	Civil Affairs, Coblenz.	1st Jan. '22 to 18 May '22.
Goodyear, Milton B.	Capt. Inf.	Serv. Col. 5th Inf.	1st Jan. '22 to 18 Mar. '22.
Gorman, Howard J.	Capt. Inf.	8th Infantry	1st Jan. '22 to 5 Apr. '22.
Graham, Ottis L.	Capt. M.C.	Antwerp, Belgium.	1st Jan. '22 to 22 May '22.
Graham, Roy C.	Capt. QMC	M.R.S. #100	1st Jan. '22 to 22 May '22.
Graham, Walter B.	Capt. Inf.	8th Infantry.	1st Jan. '22 to 7 Apr. '22.
Grant, David N. W.	Maj. M. C.	Sta. Hosp. Coblenz.	1st Jan. '22 to 7 Apr. '22.
Gray, Dell S.	Capt. D.C.	Sta. Hosp. Coblenz.	1st Jan. '22 to 16 May '22.
Greenslade, Russell M.	1st Lt. A.S.	Air Serv. Det.	1st Jan. '22 to 22 Apr. '22.

Greer, Allen J.	Lt. Col. FA	Attached 5th Inf. 1st Jan. '22 to 4 Jan. '22.
Griffin, Will C.	1st Lt. V.C.	Vet. Hq 1st Brig. 1st Jan. '22 to 16 May '22.
Hagan, Eugene E.	Capt. QMC	QM Depot & Det. Coblenz 1st Jan. '22 to date.
Hale, Harry C.	Maj. Gen. USA	Att. Hq. AFG 1st Jan. '22 to 2 Feb. '22.
Haleston, John	1st Lt. Inf.	5th Infantry 1st Jan. '22 to 9 Mar. '22.
Hall, Alex M.	Lt. Col. Inf	8th Infantry. 1st Jan. '22 to 15 Mar. '22.
Hall, John	Chap. 1st Lt.	Sta. Hosp. Coblenz. 1st Jan. '22 to 10 Apr. '22.
Hand, Harry V.	Capt. Inf.	Pt. of Antwerp 1st Jan. '22 to 5th Jan. '22. 8th Infantry. 28th Dec. '22 to 25 Jan. '23.
Haney, Harold	Capt. QMC.	QM Depot Cobl-Lutz 1st Jan. '22 to 7 Apr. '22.
Hanna, Charles W.	1st Lt. Inf.	Det. Trans. Serv. 1st Jan. '22 to 20 June '22. 8th Infantry. 2nd Jan. '23 to 25 Jan. '23.
Hanna, Lloyd M.	1st Lt. CAC	RR&C Serv. AFG 1st Jan. '22 to 30 June '22. QM. A.F.G. 1 Oct. '22 to 25 Jan. '23.
Hare, Ray M.	Capt. QMC	M.R.S. #100 1st Jan. '22 to 30 Nov. '22.
Harman, John M.	1st Lt. Engrs.	16th Engr Bn. & 1st Engrs. 1st Jan. '22 to 25 Jan. '23.
Hart, Julian G.	Capt. Inf.	5th Infantry. 1st Jan. '22 to 9 Mar. '22.
Harts, Wm. W.	Col. F.A.	Chief of Staff, AFG 31st Jul. '22 to date.
Hay, William H.	Col. G.S. (CAV)	Chief of Staff, AFG 1st Jan. '22 to 20 June '22.
Helsey, Albert B.	Capt. Inf.	Hqrs. Comd't Off. 1st Jan. '22 to 7 Apr. '22.
Herr, John K.	Maj. GS	A.C. of S.G-3 AFG 1st Jan. '22 to 8 Aug. '22.
Herwig, Hans R.W.	Maj. QMC	Off. QM Afg 1st Jan. '22 to 29 Jul. '22.
Higley, Harvey D.	Maj. F.A.	2nd Bn. 6th F.A. 1st Jan. '22 to 22nd Apr. '22.
Hill, Ployer, P.	1st Lt. A.S.	Air Serv. Det. 1st Jan. '22 to 22 May '22.
Hill, Roy A.	Maj. Inf.	1st Bn. 5th Inf. 1st Jan. '22 to 9 Mar. '22.
Hillard, Farrin A.	1st Lt. Inf.	5th Infantry, 1st Jan. '22 to 9th Mar. '22.
Hobson, George F.	Capt. QMC	Asst. Post QM Coblenz. 1st Jan. '22 to 18 May '22.

Hocker, Richard W. Capt. F.A. 2nd Bn. 6th F.A.
1st Jan. '22 to 22 Apr. '22.
Hodges, Campbell B. Maj. G.S. G-3 Sec. Hq. A.F.G.
1st Jan. '22 to 4 Jan. '22.
Hofmann, Augustine A. Maj. Inf. 5th & 8th Infantry
1st Jan. '22 to 25 Jan. '23.
Holbrook, Willard A. Jr. 1st Lt. Cav. Prov. Cav. Sq. & Hq. AFG.
1st Jan. '22 to 6 Oct. '22.
Holdridge, Herbert C. Capt. Q.M.C. M.T.S.
25 Jan. '22 to 26 May '22.
Holmes, Avery G. Capt. D.C. Sta. Hosp. Coblenz.
1st Jan. '22 to 7 Apr. '22.
Hones, William Capt. Inf. 8th Infantry
1st Jan. '22 to 25 Jan. '23.
Hopkins, F. M. Jr. 1st Lt. A.S. Air Serv. Det.
1st Jan. '22 to 16 May '22.
Hopping, Andrew D. 1st Lt. Inf. 5th Infantry.
1st Jan. '22 to 9 Mar. '22.
Hossfeld, Henry Lt. Col. Inf. Off. C. of S. Hq. AFG & G-2.
1st Jan. '22 to date.
Hough, Chester C. 1st Lt. C.E. 16th Engr. Bn.
1st Jan. '22 to 18 Mar. '22.
Hudson, Charles F. 1st Lt. Inf. 8th Infantry.
1st Jan. '22 to 5 Jan. '22.
Hunsioker, Stanley H. 1st Lt. QMC Det. Trans. Serv.
11th Feb. '22 to 18 May '22.
Hurless, Bernard F. Capt. Inf. 8th Infantry.
1st Jan. '22 to 25 Jan. '23.
Hurst, Warren C. Capt. Q.M.C. M.R.S. #100 M.T.S.
1st Jan. '22 to 18 May '22.
Huse, Gordon B. 1st Lt. V.C. Prov. Cav. Sq.
1st Jan. '22 to 7 Apr. '22.
Hutchinson, Charles G. Capt. Cav. A.D.C. C.G. AFG
1st Jan. '22 to date.
Iry, Clarence N. Capt. Engrs. 1st Engrs.
1st Jan. '22 to 18 Mar. '22.
Jackson, Stonewall Capt. Inf. 5th Infantry
1st Jan. '22 to 7th Apr. '22.
Jacques, William A. Capt. M.C. Prov. Med. Bn.
1st Jan. '22 to 7 Apr. '22.
James, Hobart H. 1st Lt. F.A. 2nd Bn. 6th F.A.
1st Jan. '22 to 22 Apr. '22.
James, Stanley L. Maj. S.C. 20th Ser. Co.
1st Jan. '22 to 2nd Feb. '22.
Jeffres, James E. Capt. Inf. 5th Infantry.
1st Jan. '22 to 9 Mar. '22.
Jenkins, Thomas G. Capt. Inf. Hqrs. Commandant's Off.
1st Jan. '22 to 22 May '22.
Johns, Glover Capt. D.C. 5th Infantry.
1st Jan. '22 to 9 Mar. '22.
Johnson, Neal C. Capt. Inf. 8th Infantry & G-1 Hq.
1st Jan. '22 to date.
Johnson, William H. Brig. Gen. U.S.A.C.G. 1st Brig. & Lia. Off. with Fr. Army.
1st Jan. '22 to date.

Joiner, Talley D.	Capt. Inf.	Co. K. 5th Inf.	1st Jan. '22 to 4 Jan. '22.
Joly, Charles L.	Maj. CWS	Off. Engr. Off. AFG	1st Jan. '22 to 18 Mar. '22.
Jones, Alfred E.	Capt. M.C.	5th Infantry.	1st Jan. '22 to 9 Mar. '22.
Jones, Glenn I.	Maj. M.C.	Sta. Hosp. Coblenz.	1st Jan. '22 to 8 Mar. '22. 22nd Apr. '22 to 20 June '22.
Jones, Thomas W.	Capt. F.A.	2nd Bn. 6th F.A.	1st Jan. '22 to 22 Apr. '22.
Jordan, Harry B.	Lt. Col. O.D.	Ord. Off. AFG	1st Jan. '22 to 30th June '22.
Keatinge, John H.	Capt. F.A.	2nd Bn. 6th F.A.	1st Jan. '22 to 16th May '22.
Keefer, Frank R.	Col. M.C.	Surgeon Hq. A.F.G.	1st Jan. '22 to 20 June '22.
Kidwell, Gaylord B.	1st Lt. QMC.	Off QM Hq. A.F.G.	1st Jan. '22 to 22 May '22.
King, Henry L.	Capt. S.C.	20th Serv. Co. S.C.	1st Jan. '22 to 7th Apr. '22.
Kingman, Allen F.	Capt. Inf.	A.A.C. of S. G-1 Hq. AFG	1st Jan. '22 to 7th Apr. '22.
Kirk, Allen F.	Capt. Inf.	Hq. Det. A.F.G.	1st Jan. '22 to 26th May '22.
Knight, Harry E.	Col. Inf.	Comdg. 5th Inf.	1st Jan. '22 to 9 Mar. '22.
Knight, O'Ferrall	1st Lt. F.A.	2nd Bn. 6th F.A. & Hq. AFG	1st Jan. '22 to 30th June '22.
Koenig, William C.	Maj. CAC	Comdr. Pt. Ant. & Hq. AFG	1st Jan. '22 to 30 July '22.
Kovarik, Joseph C.	1st Lt. Inf.	Co. F. 8th Infantry	1st Jan. '22 to 18th Mar. '22.
Krause, Emil	Capt. Inf.	8th Infantry.	1st Jan. '22 to 25 Jan. '23.
Lammons, Frank B.	Capt. Inf.	8th Infantry.	1st Jan. '22 to 16 May '22.
Laner, Walter E.	Capt. Inf.	Asst. Adj. AFG	1st Jan. '22 to date.
Leavitt, John W.	1st Lt. CAC	18th Ord. Co.	1st Jan. '22 to 26th May '22.
Lifsey, Charles Q.	1st Lt. Inf.	8th Infantry	1st Ja n. '22 to 7th Apr. '22.
Linden, Robert R.	2nd Lt. M.A.C.	Sta. Hosp. Coblenz	1st Jan. '22 to 18th May '22.
Littell, Alonzo L.	Capt. Q.M.C.	Pt. Antwerp Belgium & Hq. (AFG)	1st Jan. '22 to 30 June '22.
Lloyd, Walter K.	Maj. Chap.	8th Infantry	1st Jan. '22 to 7th Apr. '22.
Lockard, Elmer E.	Maj. F.D.	Off. Fin. Off. AFG	1st Jan. '22-31st Dec. '22.
Lodge, Henry G.	1st Lt. C.E.	1st Engrs.	1st Jan. '22 to 18th May '22.
Longstreet, James	Lt. Col. Cav.	Lia. Off Cologne	1st Jan. '22 to 18th May '22.

Lorch, Robert B.	Maj. Inf.	5th Infantry 1st Jan. '22 to 18th Mar. '22.
Lowenberg, Milton A.	Capt. QMC	Post QM Coblenz. 1st Jan. '22 to 20th June '22.
Lowry, Robert H.	Maj. M.C.	20th Serv. Co. S.C. 1st Jan. '22 to 7th Apr. '22.
Luplow, Walter D.	Capt. C.E.	Co. A. 16th Engr. Bn. 1st Jan. '22 to 18th May '22.
MacGregor, Lester E.	Capt. Inf.	5th Infantry 1st Jan. '22 to 16th May '22.
Mack, Walter A.	Capt. Inf.	8th Infantry 1st Jan. '22 to 25 Oct. '22
Magruder, Lloyd B.	Maj. IGD	Actg. Insp & Inspector AFG 1st Jan. '22 to date.
Mallon, Francis B.	Maj. Inf.	2nd Bn. 8th Infantry. 1st Jan. '22 to 9th Mar. '22.
Malone, Paul B.	1st Lt. Inf.	ADC OG 1st Brig & Hq AFG 1st Jan. '22 to 22nd Apr. '22.
Marburg, Edgar	1st Lt. CE	Co. A. 1st Engrs. 1st Jan. '22 to 31 Jan. '22.
March, Francis A. 3rd	1st Lt. F.A.	2nd Bn. 6th F.A. 1st Jan. '22 to 22nd Apr. '22.
Marsh, Fred T.	1st Lt. Inf.	5th Inf. & Hqrs. A.F.G. 1st Jan. '22 to 29 Apr. '22.
Marshall, Lawrence W.	Capt. Inf.	E&V Trng. O.Hq. AFG 1st Jan. '22 to 26th May '22.
Mayberry, Hugh T.	Capt. Inf.	5th Infantry. 1st Jan. '22 to 9th Mar. '22.
McAlpine, Harrison	Maj. Inf.	3rd Bn. 5th Infantry 1st Jan. '22 to 7th Apr. '22.
McCain, William A.	Maj. QMC	Trans. O. Hq. AFG 1st Jan. '22 to 2 Feb. '22.
McCammon, William W.	Lt. Col. Inf.	Hqrs. Commandant 1st Jan. '22 to 7th Apr. '22.
McCarthy, Leo J.	Capt. Inf.	G-3 Sec. Hq. AFG 1st Jan. '22 to 18th May '22.
McCatty, Kenneth	Capt. Cav.	Hq. Det. AFG 1st Jan. '22 to 5th Jan. '22.
McClintic, Brown S.	Maj. M.C.	Sta. Hospital, Coblenz. 1st Jan. '22 to 9th Mar. '22.
McClure, Clinton I.	Capt. F.A.-	Pt. Antwerp, Belgium 1st Jan. '22 to 20th June '22.
McGone, Alexander T.	1st Lt. Inf.	5th Infantry 1st Jan. '22 - 7th Apr. '22.
McConville, Joseph J.	Capt. QMC	Port Off Bremen 31st Jul. '22 to date.
McCormick, Thomas G.	Capt. F.A.	Am. Rep. Cologne Det. T. S. 1st Jan. '22 to 14 Feb. '22.
McCulloch, William A.	Maj. Inf.	A A C of S G-3 AFG 1st Jan. '22 to 18 May '22.
McDiarmid, Norman L.	Maj. M.C.	Med. Supply Depot. 1st Jan. '22 to 26th May '22,
McDonald, John W.	Capt. Cav.	Off. Hq. Comd't. 1st Jan. '22 to 18th May '22,

McFall, John H. 1st Lt. F.D. Off. Fin Off. (Auditor)
1st Jan. '22 to 20th June '22.

McGrath, Thomas J. Capt. QMC Off QM Hq. AFG
1st Jan. '22 to date.

McKee, Richard G. 1st Lt. Inf. Co. H. 5th Inf.
1st Jan. '22 to 9 Mar. '22.

McKeever, John W. Capt. M.C. Sta. Hosp. Coblenz.
1st Jan. '22 to 18 Mar. '22.

McKenna, Joseph W. Capt. Inf. Co. G., 8th Inf.
1st Jan. '22 to 16th May '22.

McKinney, Garfield L. Maj. M.C. Sta. Hospital Coblenz.
1st Jan. '22 to 30th June '22

McKittrick, William K. 1st Lt. Inf. Co. M. 8th Inf.
1st Jan. '22 to 16th May '22.

McLean, Henry C. Maj. Inf. Prov. Marshal Hq. AFG
1st Jan. '22-19th May '22
Lia. Off. Cologne. Hq. AFG
4th July, '22 to date.

McMahon, Bernard B. Capt. Inf. G-2 Sec. Hq. AFG
1st Jan. '22 to date.

McMahon, William C. Capt. Inf. 8th Infantry
1st Jan. '22 to 10 May '22.

McMillin, William D. Capt. Inf. Co. H. 8th Inf.
1st Jan. '22 to 19th May '22.

McNally, Edward T. 1st Lt. Chap. Chap. 8th Infantry
1st Jan. '22 to 25th Jan. '23.

McNeill, George P. Capt. M.C. 8th Infantry & Sta. Hosp.
1st Jan. '22 to 25 Jan. '23.

Mencher, Darrow 1st Lt. Cav. Disc. Barracks (Hq. Comdt)
1st Jan. '22 to 18th Mar. '22.

Merkel, Walter C. Capt. S.C. Off. Sig. Off. Hq. AFG
1st Jan. '22 to 20th June '22.

Metcalfe, Samuel L. Capt. Inf. 8th Infantry
1st Jan. '22 to 16th May '22.

Michie, Henry C. Jr. Maj. M.C. Sta. Hosp. Coblenz.
1st Jan. '22 to 1st May '22.
10th Oct. '22 to date.

Middleton, John W. 1st Lt. Inf. 8th Infantry
1st Jan. '22 to 25 Jan. '23.

Miller, Fred W. Capt. Inf. 5th Infantry
1st Jan. '22 to 9th Mar. '22.

Miller, Reuben B. Lt. Col. M.C. Sta. Hosp. & Surg. A.F.G.
1st Jan. '22 to date.

Miltenberger, Val E. Maj. M.C. Q.M. Depot Cobl-Lutz.
1st Jan. '22 to 13th May '22.

Ming, Marcus A.S. Capt. F.A. RR&C Serv. Hq. AFG
1st Jan. '22 to date.

Minus, Norman Capt. Inf. Co. C. 5th Inf.
1st Jan. '22 to 9th Mar. '22.

Minuth, Hans C. Capt. Cav. Lia. Off. Aix-La-Chap.
and Hqrs. A.F.G.
1st Jan. '22 to date.

Moore, Cecil R. Capt. C.E. Co. A. 16th Engr. Bn.
1st Jan. '22 to 18th Mar. '22

Morris, James M. Capt. Inf. Co. H. 5th Inf.
1st Jan. '22 to 9th Mar. '22.

Morrisette, James E.	Capt. J.A.G.D.	Off. J.A. & J.A. AFG	1st Jan. '22 to date.
Moss, Wentworth H.	Maj. Inf.	G-3 Section Hq. AFG	7th Jan. '22 to 30th June '22.
Mueller, Paul J.	Maj. Inf.	Amaroc News G-2 Sec. AFG	1st Jan. '22 to 30th June '22.
Muller, Walter J.	1st Lt. Inf.	8th Infantry	1st Jan. '22 to 25th Jan. '23.
Munday, Benton F.	Capt. Cav.	M.T. Co. 60 M.T.S.	1st Jan. '22 to 18th Mar. '22.
Munly, William C.	Capt. M.C.	Station Hosp.	1st Jan. '22 to 18th Mar. '22.
Munroe, Robert K.	Capt. C.E.	Off. Engr Off. Hq. AFG	1st Jan. '22 to 26th May '22.
Murphey, James A.	1st Lt. Inf.	8th Infantry	1st Jan. '22 to 1st Feb. '22.
Murphy, John L.	Capt. Inf.	8th Infantry	1st Jan, '22 to 1st Feb. '22.
Murphy, William H.	Capt. A.S.	Air Serv. Det.	1st Jan. '22 to 22 Apr. '22.
Neff, Emery, B.	Maj. M.C.	Station Hospital Coblenz.	1st Jan. '22 to date.
Nulsen, Charles K.	Maj. Inf.	Asst. Adj. AFG& 8Inf.	1st Jan. '22 to 25 Jan. '23.
O'Connell, Augustus B.	Capt. Inf.	8th Infantry	1st Jan. '22 to 5th Jan. '22.
Odom, Stanley G.	Capt. M.C.	Prov. Cav Sq. & 8th Inf.	1st Jan. '22 to 16th May '22.
O'Leary, Clarence	1st Lt. O.D.	60th Ord. Co.	1st Jan. '22 to 7th Apr. '22.
Orsinger, Remington	Capt. Inf.	Port of Antwerp	1st Jan. '22 to 20 June '22.
Osann, Herman E.	Capt. Inf.	G-2 Sec. Hq. AFG	1st Jan. '22 to 3 Nov. '22.
Owens, Charles H.	Capt. Inf.	Co. D. 5th Inf.	1st Jan. to 9th Mar. '22.
Parkinson, John L.	Maj. I.G.D.	Asst. Insp. Hq. AFG	1st Jan. '22 to 18th May '22.
Parsons, Willis B.	Capt. D.C.	Sta. Hosp. Coblenz.	1st Jan. '22 to 4 Dec. '22.
Passmore, George H.	Capt. Inf.	2nd Bn. 5th Inf.	1st Jan. '22 to 9 Mar. '22.
Patterson, Russell B.	Maj. Cav.	AAC of S G-4 Hq. AFG	1st Jan. '22 to 30th June '22.
Payne, Joseph H.	Capt. Inf.	Co. M. 8th Inf.	1st Jan. '22 to 5th Jan. '22.
Peak, Lawrence I.	1st Lt. A.S.	Mess Off. Air Ser Det.	1st Jan. '22 to 22 Apr. '22.
Peek, George M.	Maj. C.A.C.	A.C. of S. G-1	1st Jan. '22 to date.
Pegram, Edward S. Jr.	Capt. Engrs.	Co. A. 1st Engrs.	1st Jan. '22 to 5th Jan. '22.
Peters, William C.	Capt. Inf.	Co. G., 5th Inf.	1st Jan. '22 to 9th Mar. '22.
Piburn, Edwin W.	Capt. Inf.	8th Infantry	1st Jan. '22 to 18th Mar. '22.

Pickett, Jerome	Capt. Inf.	Hqrs, Comd't Off, 1st Jan. '22 to 16th May '22.
Pierce, John L.	Capt. Inf.	8th Infantry 1st Jan. '22 to 16th May '22.
Pinkerton, Hugh M.	Capt. Inf.	A.P.M. & P.M. 1st Jan. '22 to date.
Pirtle, James J.	Capt. Inf.	2nd Bn 5th Infantry 1st Jan. '22 to 25th Jan. '23.
Platt, John C. Jr.	Capt. Inf	G-4 Sec. Hq. AFG 1st Jan. '22 to 26th May '22.
Platt, Robert J.	Capt. M.C.	Station Hospital 1st Jan. '22 to 7th Apr. '22.
Powell, William D.	Capt. Inf.	Co. I, 5th Inf. 1st Jan. '22 to 7th Apr. '22.
Price, Herbert H.	Capt. M.C.	Station Hospital 1st Jan. '22 to 18th May '22.
Privett, Forbie H.	1st Lt. Inf.	8th Infantry 1st Jan. '22 to 1st Feb '22.
Proctor, Arthur B.	Capt. QMC	QM Depot Cobl-Lutz 1st Jan. '22 to 16th May '22.
Purdy, William G.	Capt. Inf.	8th Infantry 1st Jan. '22 to 25th Jan. '23.
Purnell, Harry S.	Maj. M.C.	Surg. 8th Inf. 1st Jan. '22 to 20th June '22.
Putnam, Webster F. Jr.	1st Lt. CAC	M.P. Co. 1st Jan. '22 to 18th Mar. '22.
Radford, Robert A.	Capt. C.E.	16th Engr. Bn. 1st Jan. '22 to 15 Nov. '22.
Rand, Lewis A.	Col. Engrs.	Off. Engr. Off (Att. Hqr) 1st July, '22 to 3d July '22.
Randal, Robert T.	1st Lt. Inf.	5th Infantry 1st Jan. '22 to 31st Jan. '22.
Rayens, Charles E.	Capt. Inf.	5th Infantry. 1st Jan. '22 to 9th Mar. '22.
Reasoner, Mathew A.	Maj. M.C.	Sta. Hosp. Coblenz. 1st Jan. '22 to date.
Reed, Frank F.	Capt. QMC	MRS 100 MTS 1st Jan. '22 to 16th May '22.
Reed, Ollie W.	Capt. Inf.	QM Det. Bendorf 1st Jan. '22 to 22 May '22.
Reinburg, William H. W.	Capt. Cav.	Prov. Cav. Sq & Hq AFG 1st Jan. '22 to date.
Reinhardt, Emil F.	Maj. Inf.	5th Infantry & 8th Inf. 1st Jan. '22 to 16th May '22.
Reisinger, James W.H. Jr.	Maj. Inf.	Det. Trans. Serv. 1st Jan. '22 to 31st Jan. '22.
Rice, Horace G.	Maj. F.D.	Off. Fin Off. Hq AFG 1st Jan. '22 to 26th May '22.
Rice, John K.	Capt. Inf.	Co. D. 5th Inf. 1st Jan. '22 to 7th Apr. '22.
Rickner, James C.	1st Lt. Inf.	Co. D. 8th Inf. 1st Jan. '22 to 25th Jan. '23.
Roberts, Caesar R.	Capt. CAC	MT Co. M. T. S. 1st Jan. '22 to 5th Apr. '22.
Robinson, Victor L.	1st Lt. QMC	Pt. Antwerp 1st Jan. '22 to 26th May '22.

Rodgers, John A.	1st Lt. Inf.	8th Infantry
		1st Jan. '22 to 25th Jan. '23.
Rogers, Harold D.	Capt. M.C.	Sta. Hosp. Coblenz.
		1st Jan. '22 to 16th May '22.
Rosen, Harry I.	1st Lt. Q.M.C.	Off. Q.M. Hq. AFG
		1st Jan. '22 to 22nd May '22.
Rumbough, David S.	Capt. F.A.	Hqrs. 1st Brigade
		1st Jan. '22 to 7th Apr. '22.
Sadler, William H.	Capt. Q.M.C.	Coblenz Del. Det. Trans. Serv. & Trans. Off. AFG.
		1st Jan. '22 to date.
Samford, William C.	Capt. Inf.	Co. E. 5th Inf.
		1st Jan. '22 to 31st Jan. '22.
Sanders, Augustus D.	1st Lt. Inf.	Co. B. 5th Inf.
		1st Jan. '22 to 9th Mar. '22.
Santschi, Eugene Jr.	Maj. Inf.	Att. to Hqrs. A.F.G.
		1st Jan. '22 to 18th May '22.
Schasl, William J. Jr.	Capt. Inf.	Co. D. 5th Inf.
		1st Jan. '22 to 31st Jan. '22.
Schoenfeld, Frederick	Capt. Q.M.C.	M.R.S. #100 MTS
		1st Jan. '22 to 30th June '22.
Schoonmaker, Lucas E.	Capt. CAC	G-2 Sec. Hq. AFG
		1st Jan. '22 to 18th May '22.
Schow, Robert A.	1st Lt. Inf.	8th Infantry
		1st Jan. '22 to 25th Jan, '22.
Schucker, Frank R.	Capt. Inf.	MPS #100 & 8th Inf.
		1st Jan. '22 to 30th June '22.
		2nd Oct. '22 to 25th Jan. '23.
Scott, Edwin M.	Capt. Q.M.C.	Off. Q.M.A.F.G.
		1st Jan. '22 to 31 Sept. '22.
Scott, Harold O.	Lt. Col. D.C.	Off. Surg. (Dent. Surg)
		1st Jan. '22 to 22nd May '22.
Searby, Edmund W.	1st Lt. F.A.	2nd Bn 6th F.A.
		1st Jan. '22 to 22nd Apr. '22.
Shaw, Carl R.	Capt. Engrs.	1st Engrs.
		1st Jan. '22 to 18th Mar. '22.
Sheen, Henry H.	Lt. Col. QMC	QM & AC of S G-4 Hq AFG
		1st Jan. '22 to date.
		Actg. Chief of Staff
		20th June '22 to 31st July '22.
Sherman, Harry B.	1st Lt. Inf.	Co. K, 5th Infantry
		1st Jan. '22 to 7th Apr. '22.
Shipp, Fabius B.	Capt. Cav.	Pt. Antwerp.
		1st Jan. '22 to 20th June '22.
Shipp, William E.	Maj. Cav.	RR&C O.Pt. Antwerp.
		1st Jan. '22 to 3d Feb. '22.
Shugart, Edmund R.	1st Lt. Inf.	Co. L. 5th Inf.
		1st Jan. '22 to 7th Apr. '22.
Siler, Arnold M.	1st Lt. Inf.	Mess O. Hq. Det. AFG
		1st Jan. '22 to 18th Mar. '22.
Singer, Paul L.	Capt. Cav.	Pt. Antwerp
		1st Jan. '22 to 3d Febr. '22.
Smith, Charles A. Jr.	1st Lt. Inf.	Co. I. 5th Inf & 8th Inf.
		1st Jan. '22 to 16th May '22.

Smith, Herbert S.	Maj. Chap.	Hqrs. A.F.G.	1st Jan. '22 to 10th Apr. '22.
Smith, Ross B.	Capt. Inf.	5th Inf. How. Co.	1st Jan. '22 to 9th Mar. '22.
Smith, William Alex	Capt. M.C.	Sta. Hosp. Coblenz.	1st Jan. '22 to 7th Apr. '22.
Smith, William E.	1st Lt. Inf.	3rd Bn. 5th Inf.	1st Jan. '22 to 7th Apr. '22.
Snell, Charles F.	Capt. M.C.	Station Hosp.	1st Jan. '22 to 26th May '22.
Snell, Charles F.	Capt. M.C.	Station Hosp.	1st Jan. '22 to 26th May '22.
Snidow, Robert C.	Capt. C.A.C.	[illegible] #100 MTS	1st Jan. '22 to 30th June '22.
		Off. Q.M. AFG	6th Oct. '22 to 25th Jan. '23.
Spalding, Donald P.	Capt. Inf.	Co. B. 5th Inf.	1st Jan. '22 to 9th Mar. '22.
Speidel, William H.	1st Lt. Inf.	8th Infantry	1st Jan. '22 to 16th May '22.
Stevens, Oscar G.	Capt. Inf.	C of F. 8th Inf.	1st Jan. '22 to 16th May '22.
Steward, Elwood M.S.	Capt. Inf.	Off. J.A. Hq. AFG	1st Jan. '22 to 26th May '22.
Stewart, William M.	Capt. Inf.	8th Infantry	1st Jan. '22 to 25th Jan. '23.
Stickney, Lois S.	Capt. Inf.	8th Infantry.	1st Jan. '22 to 25th Jan. '23.
Stockton, George A.	Capt. Inf.	Co. K. 8th Infantry	1st Jan. '22 to 15th Nov. '22.
Stokely, Carlin C.	Maj. Inf.	Hq. 5th Infantry	1st Jan. '22 to 7th Apr. '22.
Stokes, John H. Jr.	1st Lt. Inf.	Co. K. 8th Infantry	1st Jan. '22 to 25th Jan. '23.
Stone, David L.	Col. G.S.	LARHC & OCCA Hq. Afg.	1st Jan. '22 to date.
Strong, Alden G.	Maj. CAC	Dep. QM. & A A C of S (G-4)	1st Jan. '22 to date.
Sturges, Edward A.	Lt. Col. F.D.	Finance Officer AFG	1st Jan. '22 to date.
Sumner, Wilbur B.	Capt. F.A.	ADC CG & Sec'y Hq. AFG	1st Jan. '22 to 23 July '22.
Szilagyi, Nicholas	Capt. Inf.	Hq. 5th Inf.	1st Jan. '22 to 9th Mar. '22.
Tague, Harold M.	Capt. Inf.	ADC Atched. Hq. AFG	1st Jan. '22 to 2 Feb '22.
Talbot, Ralph Jr.	Maj. Cav.	C.O. Prov. Cav. Sq & Hq AFG	1st Jan. '22 to date.
Talbott, Samuel G.	Maj. AGD	Adjutant A.F.G.	1st Jan. '22 to date.
Taylor, Richard T.	Maj. Inf.	C.O. 3rd Bn. 8th Inf.	1st Jan. '22 to 9th Mar. '22.
Taylor, Thomas F.	Maj. Inf.	C.O. Hq. Det.	1st Jan. '22 to 23 May '22.

Thiele, Claude M.	Maj. CAC	Asst Adj. & Civ. Affairs 1st Jan. '22 to date.
Thomas, Alfred R. Jr.	Capt. M.C.	8th Infantry 1st Jan. '22 to 20th June '22.
Thomas, Thomas	Capt. Inf.	Co. E. 5th Inf. 1st Jan. '22 to 9th Mar. '22.
Thomas, William N. Jr.	Capt. Engrs.	Co. D. 1st Engrs. 1st Jan. '22 to 18th May '22.
Thompson, Richard F.	Capt. D.C.	Station Hospital, Coblenz. 1st Jan. '22 to 22 Apr. '22.
Thomson, Stanley A.	Capt. Inf.	Prov. M.P. Co. 1st Jan. '22 to 31st Jan. '22.
Thorson, Truman C.	1st Lt. Inf.	Co. E. 8th Infantry. 1st Jan. '22 to 16th May '22.
Timberlake, Edward W.	Capt. C.A.C.	M.T.C. 60 M.T.S. 1st Jan. '22 to 30th June '22.
Titus, Calvin P.	Maj. Inf.	Hq. 1st Brigade 1st Jan. '22 to 2nd Feb. '22.
Todd, Charles C.	Col. Inf.	Act'g. J.A. Hq. AFG 1st Jan. '22 to 22nd Apr. '22.
Tolson, Walter F.	Capt. M.C.	Asst. Surgeon Hq. Comdt. and Station Hospital 1st Jan. '22 to 25th Jan. '23.
Tourtillott, Raymond E.	Capt. Inf.	8th Infantry 1st Jan. '22 to 21st Jan. '22. 4th Dec. '22 to 25th Jan '23.
Tribolet, Hervey, A.	Capt. Inf.	Co. F. 5th Inf. 1st Jan '22 to 9th Mar. '22.
Tunis, Harry O.	Capt. Engrs.	16th Engrs. Bn. 1st Jan. '22 to 18th May '22.
Turner, William K.	Capt. M.C.	Station Hosp. Coblenz. 1st Jan. '22 to 9th Mar. '22.
Tyng, Francis C.	Capt. M.C.	Station Hosp. Coblenz. 1st Jan. '22 to 22nd May '22.
Van Auken, William B.	Capt. Cav.	Det. Trans Serv. 1st Jan. '22 to 13th Mar. '22.
Vermette, Raymond E.	Capt. Inf.	5th Infantry 1st Jan. '22 to 7th Apr. '22.
Wagner, Herman U.	Capt. O.D.	82nd Ord. Co. 1st Jan. '22 to 20th June '22.
Wagner, Paul S.	Capt. M.C.	5th Infantry 1st Jan. '22 to 21st Apr. '22.
Wallace Melville W.F.	Capt. Inf.	8th Infantry 1st Jan. '22 to 15th Nov. '22.
Walson, Charles M.	Maj. M.C.	Station Hospital, Coblenz. 1st Jan. '22 to 26th May '22.
Ware, Joseph F.	Maj. S.C.	Sig. Off. Hq. AFG 1st Jan. '22 to 17th Dec. '22.
Watson, William R.	Capt. Inf.	Co. C. 5th Inf. 1st Jan. '22 to 9th Mar. '22.
Wear, Hugh A.	Capt. Inf.	Co. H. 5th Inf. 1st Jan. '22 to 9th Mar. '22.
Weatherby, Francis E.	Capt. M.C.	Sta. Hosp. Coblenz. 1st Jan. '22 to 10th Apr. '22.

Webster, Louis [illegible]	1st Lt. Q.M.C.	Qm. Depot Cobl-Lutz.
		1st Jan. '22 to 16th May '22.
Weeks, Henry J.	Maj. Q.M.C.	Motor Trans. Serv.
		1st Jan. '22 to date.
Welch, John M.	Capt. M.C.	5th Infantry
		1st Jan. '22 to 18th May '22.
Wells, Everett R.	1st Lt. Sig. Co.	Off. Sig. Off. AFG
		1st Jan. '22 to 20th June '22.
West, Donald R.	1st Lt. Q.M.C.	E&V Trng. Off. Hq. AFG
		1st Jan. '22 to 31st Jan. '22.
Wheatley, Edwin T.	Capt. Inf.	8th Infantry
		1st Jan. '22 to 25th Jan. '22.
Wheeler, Richard B.	1st Lt. Inf.	Co. H. 8th Infantry.
		1st Jan. '22 to 9 Mar. '22.
Whippler, Frederic R.	1st Lt. Q.M.C.	Port of Antwerp
		1st Jan. '22 to 18th May '22.
Whisner, Emons B.	Capt. Inf.	Hqrs. Comdt. Disc. Bks.
		1st Jan. '22 to 30th June '22.
		8th Infantry.
		2nd Jan. '23 to 25th Jan. '23.
White, Henry K.	Capt. Inf.	8th Infantry.
		1st Jan. '22 to 25th Jan. '23.
Whitehead, Clayton S.	Capt. Inf.	5th Infantry
		1st Jan. '22 to 9th Mar. '22.
Whitesell, Claire A.	Capt. Inf.	Co. L. 5th Inf.
		1st Jan. '22 to 31st Jan. '22.
Wicks, Roger M.	1st Lt. F.A.	2nd Bn. 6th F.A.
		1st Jan. '22 to 22nd Apr. '22.
Willford, James W.	1st Lt. Q.M.C.	Post Q.M. Coblenz.
		1st Jan. '22 to 22nd Apr. '22.
Williams, Randolph P.	1st Lt. Engrs.	Co. B. 1st Engrs.
		1st Jan. '22 to 18th Mar. '22.
Williamson, George M. Jr.	1st Lt. Inf.	ADC Genl. Johnston.
		1st Jan. '22 to date.
Wilson, Hugh M.	1st Lt. Inf.	8th Infantry
		1st Jan. '22 to 25 Jan. '23.
Wilson, Maybin H.	1st Lt. Engrs.	16th Engrs. 1st Engrs.
		& Hqrs. A.F.G.
		1st Jan. '22 to date.
Winship, Blanton	Col. J.A.G.D.	DS Paris, Att. Hqrs. AFG
		1st Jan. '22 to 30th June '22.
Winstead, Thomas H.	1st Lt. Inf.	5th Infantry
		1st Jan. '22 to 9th Mar. '22.
Wolber, Oscar K.	Capt. Inf.	RR&C Serv. Hqrs. AFG
		1st Jan. '22 to date.
Worthington, Russel	1st Lt. QMC	Office of the QM. AFG
		13th Feb. '22 to 13th Dec. '22.
Yon, Everett M.	Capt. Inf.	Co. A. 5th Infantry
		1st Jan. '22 to 9th Mar. '22.

SUB-CHAPTER B

DISCIPLINE AND MORALE.

General Remarks.

Good discipline and high morale go hand in hand and both are essential to the successful administration of any command, especially that of an occupying force. The high standard of discipline and the excellent morale established during the early days of the Third Army and continuted in the A.F.G. during 1920-21 was maintained in 1922-23 by the constant and unhesitating employment of all the means and agencies available. The untiring efforts of the Commanding General to demonstrate to the members of the A.F.G. that Europe would judge the American nation by their conduct were successful, and it is universally recognised that the American Forces left the Rhineland with an enviable record both in the eyes of the Allies and the Germans. Naturally, the degree of discipline fluctuated from time to time, but the general trend was toward improvement. It may safely be said that the discipline of the A .F.G. in the matter of neatness, saluting and soldierly appearance, both in and out of ranks, was excellent and probably superior to that of any of the other occupying forces. When it is realised that the American private soldier in Germany received in pay more marks than the average high paid salaried German civilian and that intoxicating liquors and loose women were to be obtained easily and cheaply, it is not difficult to account for the large number of disciplinary reports. In spite of the most stringent orders faithfully executed by all officers of the command, the soldiers of the A .F. G . were subjected to temptation to an extent greater than the average other soldier in the American Army.

The following table shows by month the percentages of reports of A .W.O.L.s, disciplinary reports, courts-martial (summary and special) and new cases of veneral diseases which occurred in the whole command;

	Average strength	AWOL	Discp. reports	Courts-martial	Venereal diseases
1922					
January	7583	2.18	4.27	4.12	0.63
February	6281	2.07	2.48	3.32	0.63
March	5911	4.66	3.55	4.34	0.64
April	3880	3.51	2.91	3.48	0.50
May	2185	4.76	4.34	6.98	1.14
June	1378	2.85	7.40	5.14	0.57
July	1108	0.45	2.88	2.15	0.63
August	1088	2.20	3.85	3.21	0.82
September	1092	3.02	1.92	4.48	1.92
October	1080	2.03	2.87	3.60	0.92
November	1050	1.08	1.80	3.52	0.35
December	1059	1.41	0.66	1.88	0.75
1923					
January	836	4.18	1.31	2.87	1.31

In consulting the above table, it should be borne in mind that disciplinary reports were rendered in all instances on the least infraction of rules and regulations, such as blouse unbuttoned, hands in pockets, etc. The percentage of A.W.O.L.s during March, April and May is largely accounted for by the fact that the reduction of the forces was being carried out during these months and that many men desiring to visit their families or friends in Germany absented themselves from their organisations just prior to the sailing of transports. These absences were usually of short duration, since the absentees returned to their units in time to accompany them to the United States.

The morale of the forces remained high. Soldierly pride resulting from good discipline, contentment and confidence in their leaders, was evident in all ranks. There was a wholesome spirit of rivalry and emulation in athletics as well as in drills, maneuvres and target practice between and among the various units, and between organisations of these forces and units of the Allied Armies. Competitions of all sorts and the offering of prizes by welfare organisations to units for the best disciplinary records served to keep this spirit alive and to help maintain the high morale.

Provost Marshal's Department.

The constant reduction of the strength of the A.F.G. necessitated a corresponding reduction in both the commissioned and enlisted personnel of the Provost Marshal's Department. Early in the new year this agency consisted of five officers and 160 enlisted men and functioned under the Headquarters Commandant. The detachment of enlisted men was made up of men detailed from the various units in the A.F.G. for service with the Provost Marshal's Department.

The military prison was discontinued on April 10th, when all the enlisted personnel on detached service were returned to their organisations. All American prisoners were transferred to the 8gh Infantry Guard House and the civilian prisoners delivered to the German authorities for confinement in the German prison,

The office of the Headquarters Commandant ceased to function on April 15th, when the Provost Marshal's Department was again placed under the A.C.of S., G-1. It continued to operate under that office until the withdrawal of the A.F.G. Other reductions during the year reduced the strength of the Provost Marshal's Department to two officers and 45 enlisted men, which was its strength when the zone was turned over to the French. The officers were assigned to Headquarters, A.F.G., and the enlisted men carried on detached service from the 8th Infantry. The Department continued to operate in two distinct sections as far as personnel was concerned, namely, the Military Police and the Department of Criminal Investigation, but the reduction of the A.F.G. necessarily caused a consolidation and the interchanging of personnel to meet changed conditions. All the Military Police on outpost duty, both in the A merican area and in those areas occupied by the French and British Forces, were withdrawn to Coblenz in 1922, although the work of the Provost Marshal's Department, on a smaller scale, continued to reach into the entire occupied area of Germany as well as the interior.

The improvement in the discipline and conduct of the members of the A.F.G. led to a great reduction in the actual police work both in regard to soldiers and civilians, and, with the reduced strength of the American garrison, arrests and disciplinary reports were cut to a minimum. On the other hand, the secret police work carried on by the Department of Criminal Investigation showed no decrease. The work of the Department during 1922 covered the following heads:

General police of American area;
Enforcement of orders of Headquarters, A.F.G., and Ordinances of I.A.R.H.C.;
Apprehension and prosecution of absentees and deserters;
Apprehension and prosecution of civilians violating orders of Commanding General, A.F.G., and Ordinances of I.A.R.H.C.;
Enforcement of traffic regulations;
Investigation of character in connection with marriage applications, etc.;
Control and apprehension of prostitutes, vagrants and undesirables;
Investigations of alleged abuse.

The department recovered and returned to the government property to the value of over $8,000.00 and nearly 6,000,000 marks during the year.

Case of Staff Sgt. Collins.

Disciplinary action taken in some cases brought about situations of a complicated and international character. Because of the situation it created and the interest taken by a member of Congress in the case of staff Sergeant Collins, a resumé thereof is given below:

> Walter J. Collins enlisted at these Headquarters on December 2nd, 1919, and rose to the grade of Staff Sergeant, Air Service Detachment, A.F.G. In March, 1922, he obtained a ten-day furlough and a few days later in the same month was arrested by the Bavarian authorities, charged with the larceny and embezzlement, between the dates of November 14th and 19th, 1919, of property valued at about 12,600 marks. The Germans claimed that Collins' real name was Heidensohn and that he was a German national.

Immediately upon receipt of information at these Headquarters that the sergeant had been imprisoned by the Bavarian civil authorities, a staff officer was sent to Munich for a thorough investigation of the case and every effort was made by the Military Intelligence Division, A.F.G., to find out the true facts in the matter and to see whether, as Collins alleged, the charges were the result of mistaken identity. The officer sent to investigate was convinced that Collins was guilty of the alleged offenses. and that he was really Heidensohn, formerly a German citizen. He found the Bavarian authorities civil and apparently anxious to get rid of Collins, if possible. They offered to release him upon payment of the sum of 12,600 marks to the persons alleged to have been robbed. Collins was without funds and there was no fund available at Headquarters, A.F.G., for the payment of this amount, even had it been desired to pay it. Collins was tried by the Bavarian courts and was convicted on May 18th, 1922, under the charges, and sentenced to two and one-half months' confinement, fifteen days of which were remitted in view of the length of time he was in prison awaiting trial. Another staff officer visited Collins after the trial and talked with the American Consul and Collins himself.

It appeared from the reports of these two staff officers and the documentary proof furnished by the German authorities that Collins was identified through means of a photograph by his own mother, Frau Heidensohn, #12 Claudiusstrasse, Berlin, as being her own son and a German national. By a comparison of his own fingerprints with those in possession of Scotland Yard at London, he was shown to be a man known as Henderson, who was interned during the war as a German enemy alien. Seven confronting witnesses, the majority of whom were impartial, testified that he was the man Heidensohn, and guilty of the alleged charges. The American Consul interested himself in Collins' case to the extent of seeing that he obtained a good criminal lawyer of Munich as counsel. The first assistant in the office of the Consul, a lawyer himself, attended the trial. The Consul reported that the trial was fair and, in the opinion of this lawyer who attended the trial, Collins was guilty. Collins was therefore discharged from the service by reason of sentence imposed by civil court, in compliance with paragraph 3, War Department

Cablegram W-1027, dated July 10th, 1922.

Welfare Organisations.

All of the welfare organisations which had accompanied the Third Army into Germany had been demobilised by the beginning of the new year or had given up their activities with these forces, except the Y.M.C.A. and the Salvation Army. These organisations, with a personnel properly suited to the needs of the command, carried on welfare work along the same lines and under the same control as in the past. Almost every form of proper entertainment was provided free of charge for enlisted men by them. Through the maintenance of club rooms, writing and reading rooms, bowling alleys and restaurants, the furnishing of theatrical performances and moving pictures and through the provision of material for athletic sports, they contributed in large measure to the contentment of individuals and the maintenance of morale in the A.F.G. On account of lack of funds, decreased personnel and pressing need for work elsewhere, the commander of the Salvation Army found it necessary to withdraw its services from the A.F.G. on October 21st, 1922. All the work formerly done by this organisation was cheerfully taken over by the Y.M.C.A., which up to the dissolution of the A.F.G., was the only welfare organisation operating with the American troops.

The high qualities of the personnel of these two welfare organisations and the examples set by them, combined with their generosity and untiring efforts, were a source of great benefit to the forces as a whole and an important factor in helping to maintain discipline and morale on a high plane.

Senior Chaplain.

The Senior Chaplain continued to supervise and coordinate the activities of the Chaplains in the A.F.G., all of whom, in addition to their religious and moral duties, took an active part in the promotion of educational and athletic work in their respective organisations. In addition, the Senior Chaplain assisted in the coordination of the work done by his corps and by the welfare organisations in an effort to maintain the morale and general welfare of the enlisted personnel.

It is interesting to note that during 1922, 785 religious services were held, with an attendance of 25,449;456 marriages were performed, 32 children were baptised and 38 funerals were conducted by the Senior Chaplain and his assistants.

Comments of Distinguished Visitors.

After a visit to Coblenz, Major General Harbord, Deputy Chief of Staff, paid a tribute to the American Forces in Germany, as follows:

> "I believe the American Forces in Germany have been one of the best bodies of soldiers ever assembled in any nation. I am extremely sorry that I was unable to visit the American Rhine Forces when the A.F.G. was at its prime.
>
> "The people in the United States have

every reason to be, and are, proud of the manner in which the American Forces in Germany have represented the United States since the close of the war. I think everyone at home agrees with the declaration of General Pershing that the American Forces in Germany have been the best unit in the United States Army."

The following remarks are taken from an article published in the "Army and Navy Journal" in 1922:

"General Rogers visited the American Forces in Germany, where he made an inspection of Quartermaster Corps organisations. He was tendered an inspection and review of all the troops at Coblenz by General Allen. In a preliminary report General Rogers said in part:

'I inspected all the barracks, mess halls and kitchens, and I have never seen a more satisfactory condition. Even West Point could not excel the condition of the squad rooms'.

"Speaking of the review tendered him, General Rogers said:

'I consider General Allen's command the finest body of troops that our Army has ever produced. They are very soldierly and well set up, and military courtesy is carried out in every detail. All the transportation, both animal and motor, was in excellent condition; I have never seen better. In the march past, not a soldier in the whole command was out of step, and only one mishap occurred, one of the horses falling on account of the wet pavement. Fortunately the rider was not seriously hurt'"

<u>Officers.</u>

The first Division of the General Staff (G-1) continued to exercise general supervision over the discipline and morale of the command and over all activities concerning these matters. In discharging its responsibility for the provision of wholesome recreation and amusement for the troops, this section coordinated the military and welfare agencies engaged directly in the maintenance of discipline and morale.

In addition to Major G. M. Peck, Chief of the First Division, General Staff, the following officers served throughout the year and were directly responsible for the coordination of the agencies affecting the discipline and morale of the command:

Captain L. S. Gerow, Inf.	Athletic and Morale Officer.
Captain H. M. Pinkerton, Inf.	Provost Marshal.
First Lt. T.J.Davis, Inf.	Department of Criminal Investigation.

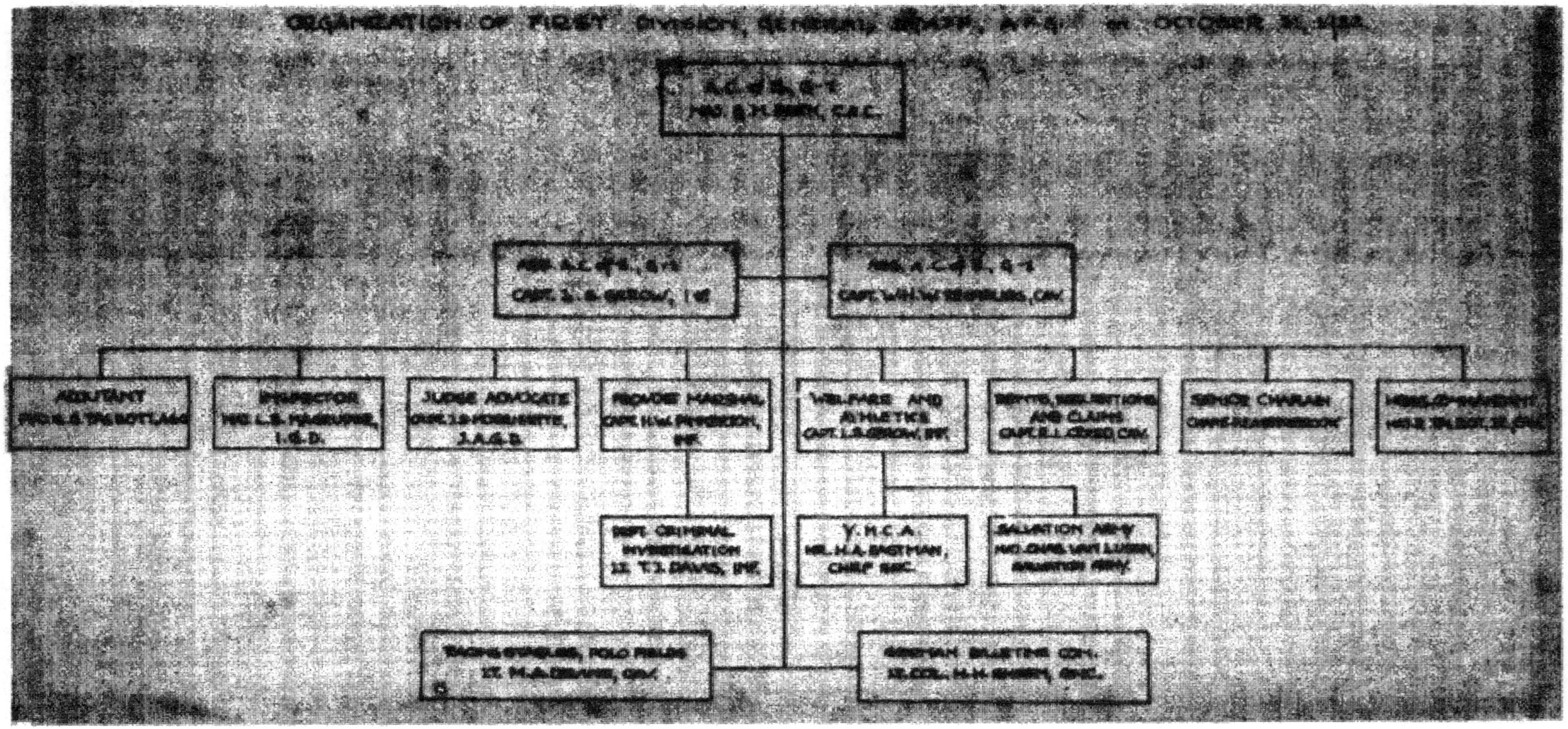
ADJUTANT
INSPECTOR
JUDGE ADVOCATE
PROVOST MARSHAL
WELFARE AND ATHLETICS
SENIOR CHAPLAIN
DEPT. CRIMINAL INVESTIGATION
Y.M.C.A.
SALVATION ARMY

SUB-CHAPTER C.

MILITARY INTELLIGENCE.

Organisation.

The general organisation of the Military Intelligence Division remained during the first half of the year practically the same as on December 31st, 1921.[1] A reorganisation then became necessary owing to the reduction of the American forces in the Rhineland. The Assistant G-2 was dispensed with and the G-2 (a-1) sub-section dealing with Enemy Order of Battle and Military Information was abolished, its greatly diminished functions being taken over by G-2 (a-2) -- Political and Economic Information. The G-2 (c) sub-section was likewise abandoned, and the supply and distribution of maps placed under the supervision of the Secretary, who was also put in charge of the Pass Bureau and the Intelligence Library. On June 20th, 1922, the day of the departure of Major F.H. Bagby, Inf., (D.O.L.) A.C. of S., G-2, for the United States, the organisation of G-2 was as follows:

A.C. of S., G-2
Secretary (Staff Sergeant)
G-2 (a-2)
G-2 (b)
Translations
Pass Bureau
Library
Amaroc News.

The strength at the time of reorganisation was

---

1 See Volume II, A.R. in O.G., 1920-21, page 290.

four officers, twenty-six enlisted men and eight civilian employees. The foregoing organisation was maintained with almost no change until January 24th, 1923, when Captain Alfred E. Dedicke and twenty-four enlisted men were released from further duty with the division in order to enable them to depart for the United States with the first contingent leaving in compliance with the order directing the withdrawal of the American Forces in Germany. Captain Herman E. Osan, who had belonged to G-2 uninterruptedly since the early days of the A.F.G., obtained his discharge from the service and left the division on November 2nd, 1922. On January 27th, 1923, when the American Forces in Germany officially ceased to exist, the Military Intelligence Division consisted of two officers. (Lt. Colonel Henry Hossfeld, Inf., A.C. of S., G-2, and Captain Bernard B. McMahon, Inf.), two enlisted men and four civilians, who were retained on duty for the purpose of terminating the affairs of the division, closing up the business of the "Amaroc News" and preparing the "History of American Representation in Occupied Germany."

Duties.

The principal mission of G-2, as heretofore, consisted in keeping the Commanding General and his staff thoroughly informed upon conditions in the American area and in procuring accurate military, political and economic information on Germany. Most of the information obtained was also forwarded to the Military Intelligence Division at Washington, the Intelligence services of the associated forces of occupation and American military attachés in Europe. In addition to the foregoing, several special tasks were assigned to G-2, the most important of which was the writing of the "History of American Representation in Occupied Germany", covering the period of 1920-21 and 1922-23. All activities were kept up until January 20th, 1923, without diminution as to scope. From that date on, a gradual reduction of the usual work became necessary in order to prepare records and maps for shipment to the United States and to turn in surplus property. Nevertheless, two bulletins were issued daily to keep the Commanding General informed on the situation in the Ruhr District and the occupied zone, which situation, since the advance of the Franco-Belgian forces on January 11th, 1923, had become critical.

Military, Political and Economic Information.

Since the final establishment of the Reichsheer (National Army) in compliance with the Versailles treaty, there was comparatively little new information available regarding the military strength and composition of the German Army. For this reason, no attempt was made to continue the periodical military bulletin after June 30th, 1922. Sufficient items of interest bearing on the German military resources to answer all purposes were supplied by the French and British forces of occupation. Valuable historical data was obtained from German sources, translated and forwarded to the Military Intelligence Division, War Department. Through the exchange of military regulations and pamphlets with the German Military Commission, Coblenz, the War Department received promptly first-hand military information, which otherwise would have taken months to secure.

The political and economic events were closely followed by G-2 (A-2) and published in bulletin form as in past years. The daily review of the German press was continued without change until September, 1922, when it was supplemented by excerpts from the French and British press and the American papers published at Paris. With this innovation, the average reader wa

good estimate of prevailing con

to other newspapers.

Secret Service.

The customary functions of this sub-section due to the lack of adequate personnel after the reduction in June, 1922, had to be considerably curtailed. There were only four Intelligence Police on duty, with one officer in charge. After abolition of certain duties, the necessity for which no longer existed on account of changed conditions, the following remained and were efficiently carried out:

Observation of civil population;
Attending and reporting upon political meetings;
Procuring military, political and economic information;
Watching industrial controversies;

Detecting and thwarting improper propaganda;
Liaison and cooperation with similar agencies of associated armies;
Liaison and cooperation with Provost Marshal, A.F.G.

Civil Affairs and G-4 were given early notice whenever there were indications that a strike might affect public utilities, and several labor troubles were averted through this close cooperation.

Translations.

There was no change in the activities of this sub-section, a closer description of which may be found in the history of 1920-1921.[1] An average of 40,000 words was translated each month in addition to the routine work for G-2. Several very important articles dealing with artillery inventions were prepared for the Ordnance Department. The decrease in the work accomplished was due to the general reduction of these forces, which also affected the translations personnel.

Maps, Pass Bureau and Library.

A considerable supply of maps was inherited from the Third Army G-2 (c), and a further lot procured by purchase in 1921. With the approaching withdrawal of the American forces from the Rhineland, it became necessary to prepare the maps for shipment to G-2, War Department, which had asked for them. 157 boxes of maps were shipped on board the "St. Mihiel" on January 25th, 1923; the remainder was sent later.

---

[1] Volume II, A.R. in O.G., p. 299

The Pass Bureau, established on January 10th, 1920, continued its activities in the usual manner under supervision of the Secretary. Although this office attended primarily to members of the A.F.G. who desired visas to pass through Germany, it operated also for the benefit of German nationals desiring a circulation stamp permitting them to enter the Saar District. Besides, members of the German Army on furlough in the American area were required to report to it in person and submit their papers for inspection. The Pass Bureau also frequently lent assistance to American civilians applying for information. Close liaison was maintained with the German Pass Office of Coblenz, an establishment of the German Foreign Office and an adjunct to the Office of the German National Commissioner attached to the Interallied Rhineland High Commission. The German authorities, as a courtesy to the American forces of occupation, agreed to visa passports or emergency passports for bona fide members of the A.F.G. desiring to visit unoccupied Germany, for $.35 and $.70, according to circumstances in lieu of the $10.00 fee usually exacted from American citizens. It was primarily due to the efforts of the official in charge, Legationsrat von Levetzow, and his assistants, that there was a complete absence of friction in the various transactions arising. The officials of the German pass office did their utmost to please -- to no small degree, perhaps, for the purpose of counteracting the bad impression created by the

Bavarian authorities, who had become notorious in throwing obstacles in the way of travellers desirous of visiting their country.

The Library, which was received during the early days of the occupation through the agency of G-2, War Department, served its purpose well. There were many calls for books dealing with the world war and the Franco-Prussian war, which officers studied prior to going on a battlefield trip. Very few accessions were made during the year, no funds being available for purchases.

"Amaroc News."

The "Amaroc News" continued publication throughout 1922 and up to, and including, January 24th, 1923. Beginning with a press-run of 4,935 in January, 1922 (due to troop sailings in December, 1921, and January, 1922, about 1,500 subscribers had been lost from the preceding month), it steadily lost in circulation through the constantly decreasing strength of effectives of the A.F.G. The personnel was cut accordingly, soldiers being replaced by Germans whenever possible.

The circulation fell off throughout the year, as is shown by the following figures:

Jan.,	1922	4935	May,	1922	3028	Sept.,	1922	1687
Feb.,	1922	4903	June,	1922	2125	Oct.,	1922	1672
March,	1922	4450	July,	1922	1883	Nov.,	1922	1555
Apr.,	1922	3543	Aug.,	1922	1847	Dec.,	1922	1563
						Jan.,	1923	1485

All these losses were caused by the evacuation of excess personnel of the A.F.G. to the States.

Plans had been made in March for the discontinuance of publication of the "Amaroc" on May 31st, since according to orders, the main body of the A.F.G. was to have left Europe by that date. In May, however, the Commanding General decided to have the paper maintained until June 15th. In June, when orders were received for the retention of approximately 1,200 officers and men, the paper had practically to begin anew, since all advertising and subscription contracts had been broken off and the features of the paper discontinued. Energetic efforts resulted in procuring an average minimum of 100 centimetres of advertising per day, and in signing up practically every soldier and civilian American in the area as a subscriber. In addition, many of the personnel of the embassies and consulates in Europe, as well as libraries and former A.F.G. members and welfare workers in the States, were induced to subscribe. Thus, by the end of June, the paper was almost self-supporting.

Due to the fall of the mark and the increase in operating costs, it was found necessary to raise the price per copy from three to five marks and to increase the price of advertising by 50% in July. The greater part of the cost of the Ehrenbreitstein fire-works exhibition on July 4th was borne by the "Amaroc", which paid out more than $100.00 therefor.

The news furnished the "Amaroc" by the Paris edition of the "Chicago Tribune", while not very satisfactory, was costing $100.00 per month, in addition to telephone tolls for the nightly talks to Paris. Arrangements were made in August for the reduction of this charge to $75.00, and also for payment by the government of the ration allowances for the enlisted men on the paper. Further arrangements were made in November for discontinuance of the "Chicago Tribune" news service and the use of the "New York Herald" service, which was obtained for $30.00 a month. This service, while much cheaper in cost, proved to be fuller in its scope and much more satisfactory in every way than that furnished by the "Tribune." In the same month, the feature services of the International News Service and the Premier Syndicate were procured gratis.

The price per copy had to be raised from time to time in keeping with the advancing costs. It was raised from five to ten marks in August; on October 12th, to twenty marks; on October 28th, to thirty marks and, on November 9th, the permission of the Commanding General was received to raise it to the daily equivalent (in marks) of one cent. This practice was continued up to the end, and the price was found ample to insure the continuance of publication without loss.

The last issue of the paper was published on the morning of January 24th, 1923[1]. After the paper had gone to press, all of the men of the staff, with the exception of

---

[1] Following liquidation of the Amaroc property, Captain McMahon turned remaining funds (4,000,000 marks, about $200.00) over to Burgomaster of Coblenz for purchase of milk for poor children.

one man retained for liquidation purposes, reported to the organizations to which they had been assigned, and left Coblenz in the afternoon to sail for the States on the "St. Mihiel."

The personnel of the staff, which in January, 1922, consisted of one officer and thirty men, had been composed of one officer and eighteen men since June, 1923. Major Paul J. Mueller, Infantry, who had been editor and manager since October, 1920, left Coblenz in June, 1922, for the States, his place being taken by Captain B.B. McMahon, Infantry, who continued as editor and manager until the last issue.

The German Military Commission.[1]

The strength of the German Military Commission (Deutsche Militaer-Kommission) on January 1st, 1923, was two officers and six enlisted men. There was practically no change in the Commission's functions, which continued to be mainly of an informatory nature, throughout 1922. The most important services rendered by this body consisted in furnishing information on military, political, economic, charitable and other organizations; information regarding missing soldiers and deserters and information pertaining to marriages between American soldiers and German nationals. The exchange of service regulations and other professional matters of a non-confidential character

---

1 Origin of Commission, see Vol. II A.F. in O.G. 1920-21, page 305.

was continued as in previous years, with the approval of G-2, War Department. The Military Commission succeeded in procuring several historical essays and accounts of battles and engagements of the world war in which American troops took part. Besides this, Major Erich Kuehlenthal, Chief of the Commission, acted as an intermediary between the German Ministry of War and the A.F.G. whenever the necessity for such direct liaison arose.

Relations between the German Military Commission and the American General Staff were always correct, and no indications of improper activities on the part of the Germans were ever observed. Major Kuehlenthal was relieved by his government on January 26th, 1923, after four years' association with these forces. He displayed marked ability, discretion and tact while occupying the position as Chief of the German Military Commission, and received the personal commendation of General Allen, at the time of his departure, for the manner in which he had discharged a difficult task. The other officer on duty with the Commission was First Lieutenant Helmer.

Liaison.

There was, as in the past, constant liaison with the Military Intelligence Division, War Department, with the Intelligence services of the other forces of occupation and with most of the American military attachés in Europe. All publications of interest were sent them, with the exception of G-2, War Department, which office was furnished not only the prescribed reports but also many other translations, books and documents of military value. The relations with the Intelligence services of the associated armies of occupation continued to be most cordial, occasional conferences being held for the purpose of exchanging information. In addition to forwarding the regular G-2 publications to American military attaches, the customary circular weekly letter containing information on Germany and items of Army interest was sent out, the last one being dated January 23rd, 1923.

Officers.

Major P. H. Bagby, Infantry (D.O.L.)*, who had served as A.C. of S., G-2, since the creation of the A.F.G., left for the United States on June 20th, 1922, in compliance with War Department instructions. Lieut. Colonel Henry Hossfeld, Infantry (D.O.L.)* succeeded Major Bagby as Acting A.C. of S., G-2, and served in that capacity until January 8th, 1923, when he was appointed A.C. of S., G-2 A.F.G.

The following officers served in G-2 during some portion of the period January, 1922-January, 1923 :

G-2 (a-1)

Capt. B.B. McMahon, Inf., (until June, 1922)

G-2 (a-2)

Capt. B.E. Schoonmaker, C.A.G. (D.O.L.)
Capt. A.E. Dedecke Inf., (D.O.L.)

G-2 (b)

Capt. H.E. Osann, Infantry*

Translations

---

* Officers who served in G-2 for more than 3 years.

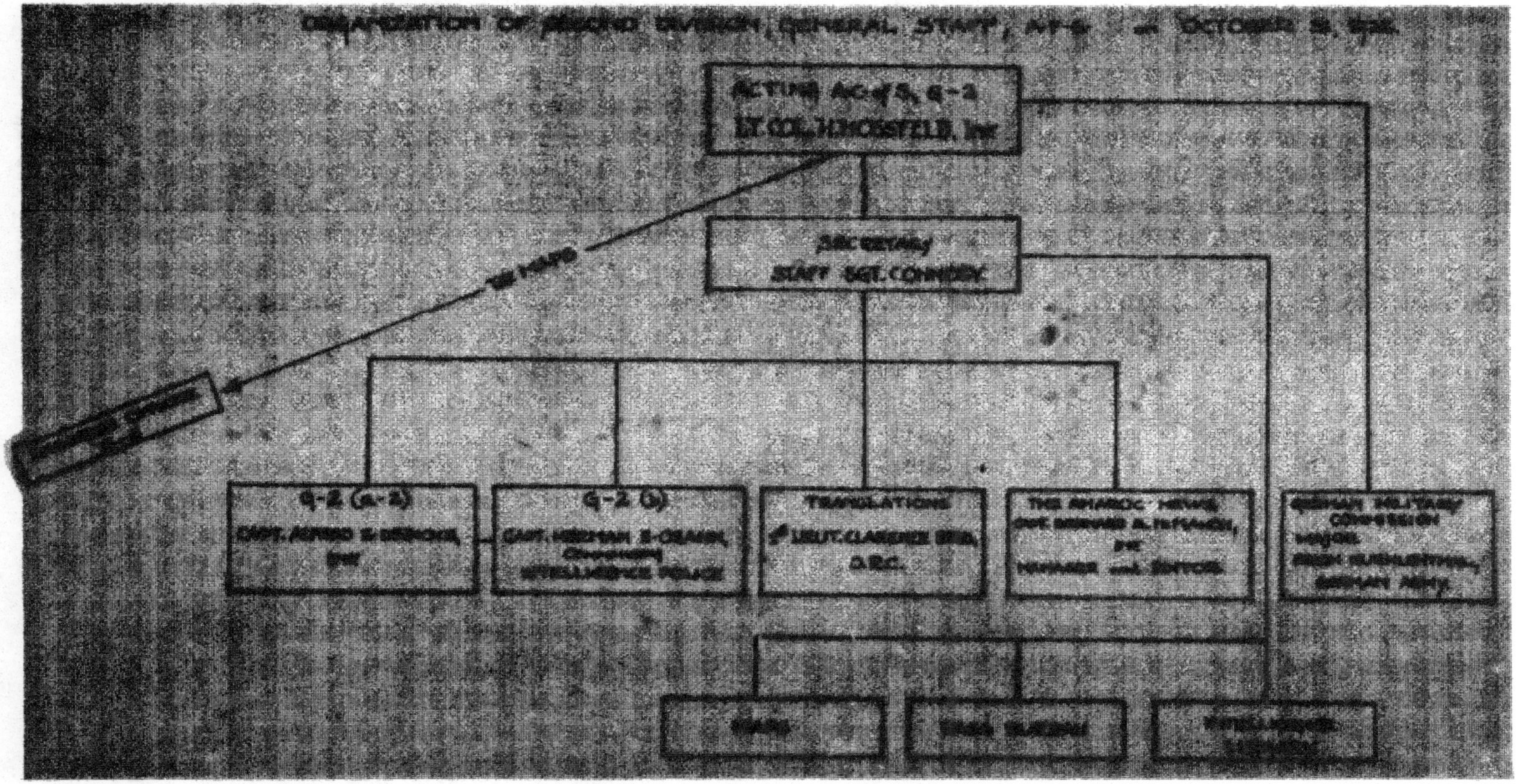
ACTING AC of S, G-2
SECRETARY
G-2 (a-2)
G-2 (b)
TRANSLATIONS
O.R.C.
MANAGER and EDITOR
GERMAN MILITARY COMMISSION
GERMAN ARMY

1st Lieut. C.W. Bird, O.R.C.

Amaroc News

Major Paul J. Mueller, Inf. (Jan. 1st, 1923, to June 15th, 1922)
Capt. B.B. McMahon, Inf. (June 15th, 1922, to Jan. 24th 1923)
(Staff Sergeant William H. Connolly served as Secretary of G-2 for more than six months)

Combat Intelligence.

The training of the Intelligence services of the various large units was supervised closely until the reduction of the American forces on the Rhine in June, 1922, and the high standard reached by the troops in that respect was maintained. When, after the final reduction, only two battalions of the 8th Infantry remained, no further effort was made in this direction, since the Regimental Intelligence Officer was perfectly capable of attending to the supervision of the Intelligence training. The Manoeuvres in 1922 proved that the troops were well instructed in Combat Intelligence.

---- ooo O ooo ----

SUB-CHAPTER D.

MILITARY TRAINING AND OTHER ACTIVITIES. THIRD DIVISION, GENERAL STAFF.

General.

The policy of training instituted by the Commanding General early in the existence of the A.F.G. produced a well-balanced command in a high state of general fitness for active service in 1920 and 1921. Not only was it an efficient combat unit in general, but in such matters as military bearing and neatness of appearance, military ceremonies, equitation, sports and athletics it had acquired in Europe a high reputation for all-round efficiency.

The beginning of the new year saw these forces again following a thorough program of training preparatory to the more intensive training of spring, summer and autumn. With the reductions of the first few months of 1922, all the air forces and artillery units were returned to the United States, and it was evident that the combat exercises and manoeuvres with live ammunition, involving the use of all arms, which had characterised the end of the 1921 training season, could not be repeated in 1922 on such an extensive scale. In order that the same standard of excellence might be maintained with the reduced forces, training programs had to be modified to meet the new conditions, with a view to the development of the material at hand into the best possible composite unit.

Military Training.

Tactical.

The doctrine, principles and methods set forth in Training Regulations No. 10-5, War Department, 1921, were used as a guide, and all officers were required to familiarise themselves therewith. Schools for officers, non-commissioned officers and privates were conducted throughout the school year. These included basic courses for officers and non-commissioned officers in administration and operation of the various arms and services. The language and equitation classes were continued and attendance by all officers was compulsory. The officers' schools were conducted as prescribed by Special Regulations for Officers in Unit Schools, War Department, 1922, and were attended by officers on staff duty as well as those on duty with troops.

When, in the spring of the year, instructions were issued for the withdrawal of the A.F.G., all training activities were suspended to allow sufficient time for turning in property not to be taken to the United States and for the preparations incident to the total evacuation of the forces. As a consequence, the range work of all troops was delayed and was not completed until the first week in September. Notwithstanding this interruption, the results obtained were excellent.

The second week of September was devoted to the small arms competition and the third to the dismantling of the ranges and preparation for the manoeuvres near Bassenheim. These took place in the latter part of the month and included the solution of tactical problems by battalions supported by one-pounders and stokes mortars. They brought into play all the pahses of tactical training permitted by the reduced forces.

October was devoted to the correction of deficiencies noted during manoeuvres and the preparation by officers of map problems and field exercises. November and December were devoted to attainment of proficiency in all individual instruction except markemanship. In addition, schools for officers, non-commissioned officers and selected privates were maintained up to the receipt of the orders on January 11th, 1923, for the withdrawal of the A.F.G.

Plan of Defence.

After the strength of the command was reduced to a force not to exceed 1,200 officers and man, it became advisable to have the A.F.G. reinforced by French troops. The French garrison consisted of the 151st and 156th Infantry and the 230th Field Artillery (one battalion of 75's and one battalion of 155's). Since these troops were under the command of the Commanding General, A.F.G., and formed a part of the forces holding the Coblenz bridgehead, it became necessary to change the plan of defence for the American area. This was done in order to meet the new needs of the situation.

Staff Ride.

Through the courtesy of the French Ministry of War and General Degoutte, Commanding General, Allied Armies of Occupation, the Commanding General was authorised to send certain of the A.F.G. officers on staff rides during the school year 1922. These rides, as planned by the G-3 Division, included a tour of the battlefields of Metz (war of 1870-71). Verdun and that portion of the Meuse-Argonne sector fought over by American troops. Through the courtesy of the French, the American officers were conducted over the battlefields of Metz and Verdun by an officer detailed from the French General Staff, who delivered interesting and instructive lectures on the fighting around these places. Although two of these trips were planned, only one was actually made. This was due to the withdrawal of the A.F.G. The one trip that was made fell between December 4th and 11th, and included about twenty officers, who traveled in machine-gun trucks furnished by the A.F.G.

Equitation.

Following the custom already established, classes in equitation were held three times weekly from October until receipt of the order for withdrawal. With the reduced commissioned strength, it was possible to mount every officer in the command, and all were required to undergo this instruction unless excused for physical reasons. The results achieved were wholly satisfactory.

Athletics.

Athletics of all sorts were always developed to

a high degree in the A.F.G. Following the reduction in the spring of the year, the athletic program was necessarily reduced to an extent commensurate with the needs of the remaining forces. Athletic programs were drawn up with a view of reaching as many men as possible and for the purpose of preparing teams for Interallied competitions. Much attention was given to mass games in which all men were required to participate, as well as to standard sports such as baseball, football, basketball, swimming, golf, tennis, soccer, boxing, wrestling and track athletics.

The athletic Department of the Y.M.C.A. continued its very liberal policy in furnishing funds, equipment and personnel for carrying out the athletic programs prescribed. These programs were so arranged as to provide athletics for every individual, first, by offering a large number of varied standard sports and a wide choice, and second, by compulsory participation in mass games, which were included in the regular military training schedules.

Interallied Competitions.

In the Interallied competitions held throughout the year, in which teams from the four occupying armies participated, the success of the American representatives was always marked.

At the Interallied small arms competition (rifle - automatic rifle - machine-gun and pistol) held by the French Army at Dreve, Germany, on November 9th, the American team won first place in both the officers' and enlisted men's class.

At the small arms competition held by the British Army of the Rhine at Urbach, Germany, October 3-6, the Americans won first place in the rifle team event and in the machine-gun competition.

Interallied horse show, Coblenz, May 26-27: Americans won officers' jumping event (also second, third, fourth and fifth places). There were forty-six entries.

Concours hippique, Mayence, June 8-15: Americans won three events and took second place in four others.

British Army of the Rhine horse show, Cologne, September 20-21: In the championship jumping, Americans took second place. In the Interallied team jumping (each team consisting of three officers), the American team won the event.

Polo.

The polo players and ponies entered upon the 1922 season in excellent form, due to the long play of the season of 1921. The A.F.G. polo team visited Cologne and, playing with the 14th Hussars, the strongest of British teams, won both games scheduled with that organisation. In the interallied polo tournaments held at Coblenz on June 26th and July 23rd, 1922, the American teams won first place.

Racing.

The monthly race meets begun in 1921 were held throughout the season of 1922 on the A.F.G. race-course around the polo field. In contesting with the British and

Germany, and these forces were represented by especially qualified officers at the headquarters of each of the Allied Armies.

The functions of the American Liaison officers underwent no change during the year. They kept in close touch with the staff of the army to which accredited, were informed, and kept American Headquarters posted, on all matters of mutual interest, and were the usual agencies for the communication of routine matters. In addition to their purely military duties, they rendered assistance to American travelers, civilian as well as military.

The following list shows the American liaison officers on duty with Headquarters of the various Allied Armies:

		Assigned:	Relieved:	
	Minuth, Capt.H.C.,Cav.	Dec.2,1920.	On withdraw- of AFG.	Belgian Army.
*	Bartlett, Major C.G.,Inf.	Feb.2,1920.	Aug.1,1922.	DGCRA
*	Caffery, Major C.S.,Inf.	Aug.1,1922.	On withdraw- of AFG.	DGCRA
	Coleman,Col.LeV.,CAC.	Oct.31,1919.	June 14,1922.	French Army.
	Johnston,Brig.Gen.Wm.H.	June 16,1922.	On withdraw-al of AFG.	Allied High Command
	Longstreet,Lt.Col.James,Cav.	Jan.10,1921.	May 6,1922.	British Army.
	McLean,Major H.B.,Inf.	July 6,1922.	On withdraw-al of AFG.	British Army.

* The duty performed by these officers fell more under the Fourth than under the Third Section, General Staff.

The following Liaison officers from the Allied Armies were on duty with Headquarters, A.F.G., during 1922, and until the withdrawal of the American Forces:

BRITISH MISSION.

Jeffreys, J. Lt. Colonel.

FRENCH MISSION.

Hendrickx, P. Captain.

BELGIAN MISSION.

DeHase, Jos.Moans. Captain.

Officers.

The following officers were on duty with the Third Division, General Staff, for the period indicated:

Name:	Assigned:	Relieved:	Duties:
Herr, Major J.K., G.S.	Jan.27,1920.	July 26,1922.	ACofS.G-3.
McCarthy, Capt.L.J., Inf.	June 16,1921.	May 6,1922.	Asst.
McCulloch, Maj.W.A., Inf.	Sept.11,1921.	May 6,1922.	Asst.
Downer, Maj.J.W., F.A.	July.25,1921.	On withdrawal of AFG.	Act.AC of S.G-3.

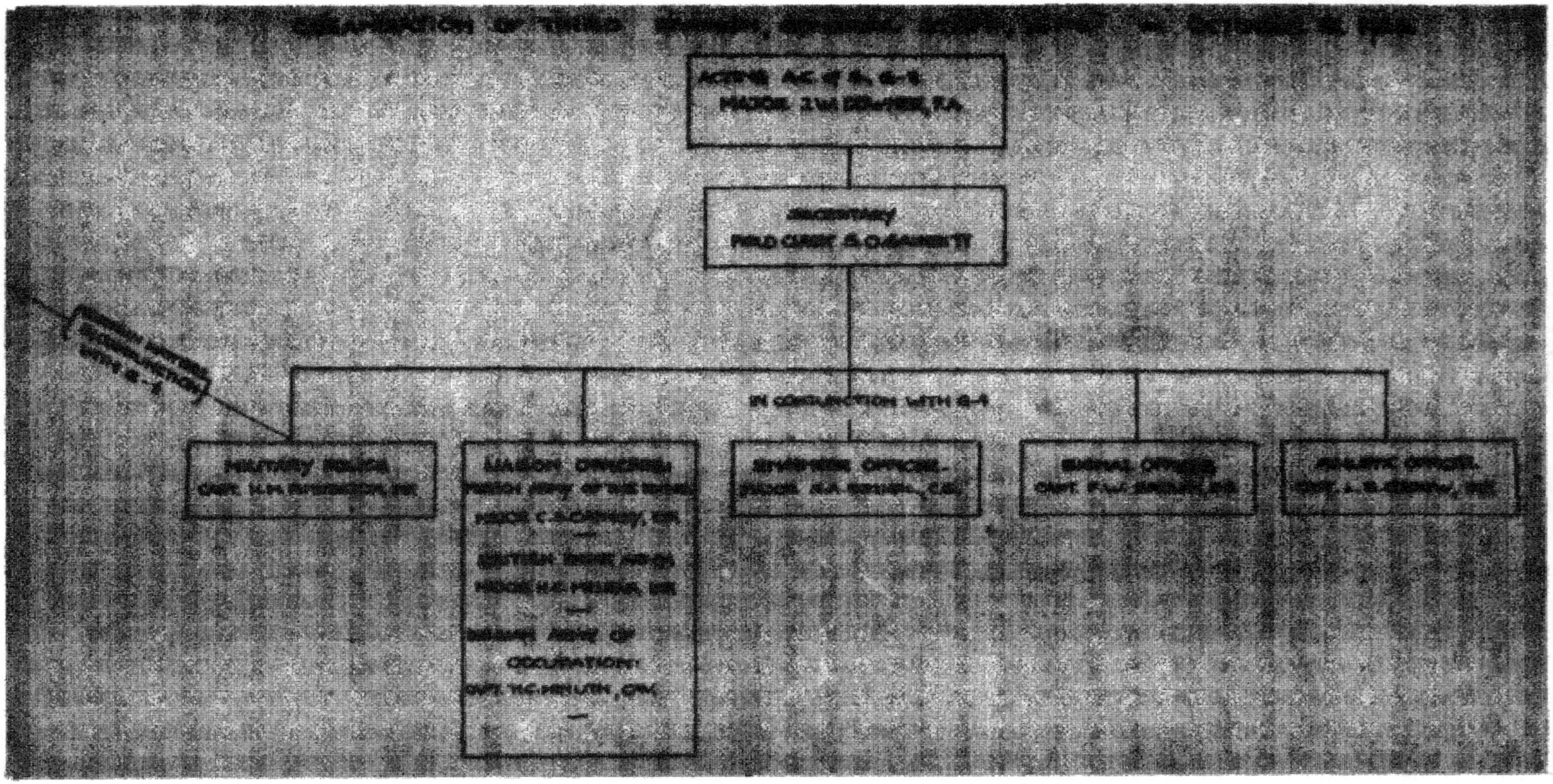

SECRETARY
IN CONJUNCTION WITH G-4
MILITARY POLICE
LIAISON OFFICERS
ENGINEER OFFICER
SIGNAL OFFICER
ATHLETIC OFFICER
OCCUPATION

SUB-CHAPTER E.

TELEPHONE AND TELEGRAPH COMMUNICATION.

Signal Communications under Reduced Forces.

The Signal Service, A.F.G., continued to supervise and control the maintenance and operation of all the Signal Corps activities throughout 1922. Considerable experimental work was carried out in an attempt to organise radio-telephony to be operated between Coblenz and Antwerp. However, due to the uncertainty of the retention of the troops on the Rhine and to the cessation of our activities at Antwerp, these experiments were abandoned.

The reduction of personnel in the American forces and the corresponding reduction in the Signal troops necessitated abandonment of the telephone and telegraph test stations at Thionville,(France), at Andernach, Montabaur, Bendorf and Neuwied,(Germany); and the photographic laboratory at Coblenz. The 2nd Signal Company, which had been on duty with the A.F.G., was reorganized into the 20th Service Company, Signal Corps, and the personnel was gradually reduced during the year from six officers and 150 men to one officer and thirty-five men, which force continued on duty until the recall of the American troops.

Employment of German Telephone Operators.

Along with this reduction, it became necessary

to employ German female telephone operators in order that the twenty-four hour telephone service should not suffer. It is interesting to recall that during the days of the Third Army, when the American troops operated and controlled hundreds of miles of lines of all sorts, the telephone operators were American girls sent into Germany from the Signal Corps units of the A.E.F. They disappeared with the Third Army, and enlisted personnel was used at the switch-boards during 1920-21.

The reduction in the strength of the A.F.G. did not proportionately reduce the amount of traffic over the telephone wires. This is shown by the fact that there was an average of 7,500 to 8,000 calls per day during the latter months of 1921, while they averaged about 5,000 per day in December, 1922.

Cables and Radio.

Arrangements were made to route cables through Paris and London, and radiograms were dispatched by wire from Coblenz to Paris and Lyons, and, from Lyons, to the United States by radio.

With the closing of the Port of Antwerp the radio station at that point was withdrawn and one of the stations at Coblenz was closed. Every effort was made to return to the Germans all lines which were not urgently needed for business of the American occupation. A great deal of long-distance traffic was routed through the German central office during the latter part of the year.

Maintenance of Equipment.

Operating under the Signal Service, A.F.G., was the Signal Corps Supply Depot, which furnished all Signal Corps supplies for the troops in the area. These supplies were inspected and the personnel in the various organisations was instructed by Signal Corps men. This system, in conjunction with proper instruction of operating personnel within each organisation, assured efficient service and served to maintain equipment in good condition.

Officers.

The following list shows the commissioned personnel on duty with the Signal Service, A.F.G. during the calendar year:

Major J.F.Ware, Signal Officer, Jan.1,1922 - Dec.15,1922.
Major Stanley L. James, Jan.1,1922 - May 16,1922.
Captain Walter C. Morkel, Jan.1,1922 - June 19,1922.
Captain H. L. D. King, Jan.1,1922 - April 6,1922.
First Lt. E. R. Wells, Jan.1,1922 - June 19,1922.
Captain Frank W. Brown, Signal Officer, Dec.16,1922 until withdrawal.

SUB-CHAPTER F

SUPPLY AND TRANSPORTATION.

Supervision.

The Fourth Division, General Staff, continued to supervise and coordinate the activities of the Quartermaster Corps, Engineer Corps, Ordnance Department, Medical Department, Finance Department and, as far as supplies were concerned, the Signal Corps. The American Mission with the D.G.C.R.A. and the American delegates with the Interallied Railway Commission and Interallied Waterways Commission also came under this division. The activities of the Engineer and Signal Corps, the Ordnance, Medical and Finance Department and the various Interallied Commissions are treated in detail in other chapters. This one deals mainly with questions of supply and transportation directly affecting the services coordinated by the Fourth Section.

Concentration.

Coincident with the reduction in strength of these forces, effort was made to concentrate all their activities. The Quartermaster Corps released its depot and barracks at Coblenz-Luetzel, as well as the salvage and remount depots, for occupancy by French troops. These activities were concentrated at Coblenz and in the Motor Transport compound. The Ordnance Depot at Metternich was abandoned and all general ordnance activities were concentrated at the Muelheim ammunition depot. Late in the year, fifty-seven of the buildings at this depot were turned over to French troops and all A.F.G. Ordnance activities concentrated in ten buildings at the west end of the depot. The Medical Department Training Centre at Neuwied was given up and the activities formerly performed by that organization were transferred to the Station Hospital. A large portion of the Station Hospital was turned over for use by the French. Engineer and Signal Corps activities were likewise reduced and concentrated whenever possible. No concentration was necessary in the Finance Department.

The reduction of the American forces to a number below 1,200, the arrival of about four times that number of French troops in the area, French occupation of Rhenish territory to the southeast and north of the American sector greatly exceeding in extent that occupied by these forces - all this necessitated strict adherence to the established policy. That policy provided that on all Interallied matters pertaining to the Fourth Section and affecting the American area, authorization of American Headquarters be obtained prior to issuing any orders applicable to that area.

Supplies.

General.

The two main sources of food supplies - requisition from the United States and purchase or requisition in Europe - were not altered throughout the period covered. Reduction in strength, discontinuance of transport service and improved economic conditions in Germany, however, brought about certain modifications.

With decreased activities in all services after June, the amount of supplies used by these services was decreased in proportion. The stocks on hand in the Medical Supply Depot, for instance, augmented by a few purchases on European markets, were sufficient to serve the needs of the Medical Department; supplies and equipment for the Engineer Corps did not require replenishing; Ordnance and Signal supplies were more than sufficient; most of the supplies coming from the Quartermaster Corps were sufficient to cover the needs of the reduced forces. Requisitions on the United States were thus reduced to a minimum.

Food.

The A.F.G. never procured food supplies by direct requisition upon the inhabitants, and not until late in 1920 was food even purchased locally to any extent. Economic conditions improved gradually during 1921, and, coincident with the cessation of the Army transport service to Europe, reached a point at which local purchases of food did not involve hardships on the inhabitants of the occupied territory. The practice begun in August, 1920, was therefore extended and the policy was adopted of purchasing or requisitioning food supplies locally whenever possible instead of obtaining them by requisition from the United States. It should be remembered that in the application of this policy, all requisitions were made on the proper German government officials, who procured the supplies in the best manner possible, thus avoiding any interference on the part of the Army with the economic progress of the occupied territory.

Food supplies thus obtained included fresh vegetables, such as potatoes, cabbage, beets and others in season; fresh meats and other components of the ration adopted to meet the needs of troops serving in Europe. Arrangements were made with the German government in November whereby even the fresh beef used by the American troops was obtained by purchase on local markets.

Fuel.

The supply of coal in Germany since the armistice was never more than moderately satisfactory. Indeed, on several occasions, the basic shortage of this essential article, supplemented by transportation and labor difficulties, made the continued supply of the American area a matter of doubt. This situation necessitated the adoption of some method in the first stages of the occupation to ensure an uninterrupted supply sufficient for the needs of the American troops and the public utilities in the American zone. This method involved the seizure and requisition, from Rhine barges, of coal destined to points in unoccupied territory. This was an effective weapon and was used only in cases of emergency.

There was no change in the manner of procurement of fuel in 1922, and only once was it necessary to resort to extreme measures, i.e., seizure. Wood in sufficient quantities was purchased from German contractors direct, or was requisitioned through proper German governmental agencies. Kerosene, gasoline, motor oils and greases were likewise obtained from German sources.

Clothing.

The procurement and supply of clothing was continued under practically the same system as in the United States. The need of various minor articles in quantities insufficient to warrant a requisition from the United States occasionally necessitated purchases locally.

Forage.

This was obtained locally, either by purchase or through requistion on the German government in the same manner as were certain articles of food and clothing.

Animals..

Other than losses throgh normal death rate, the animal strength of the A.F.G. underwent no change until after the receipt of the order in January, 1923, to withdraw the forces. Soon thereafter, approximately forty selected mounts, including some privately owned, were shipped to the United States through Antwerp and eighteen others were sent to France for the use of various U.S. officers attending schools at or near Paris. Pursuant to War Department instructions, all other animals were transferred to the Veterans Bureau, the Department of Agriculture, or were declared surplus and sold in Germany by the General Sales Agent.

Quartermaster Farm.

This farm, established near Muelheim in 1920, was originally used as an agricultural school for the A.F.G. As such, it offered courses in animal husbandry, dairy husbandry, gardening and general agriculture. With the cessation of the educational and vocational training activities and the passing of the farm to the Chief Quartermaster, it became a rather important source of food supply to the American Forces. Among the articles produced by it were fresh meats (except beef), poultry, vegetables, eggs and milk. The supply of dairy products assisted materially in the solution of the problem of obtaining sufficient quantities of good milk for adult sick and children of the Allied colony of Coblenz.

General Sales Store.

Although the volume of business done at the General Sales Store decreased greatly during the year, it continued to operate as a very excellent general establishment supplying practically all the wants of those to shom the privilege of purchase was extended. It maintained a grocery department, butcher shop, tailor shop and candy and tobacco centers. All articles of an officer's equipment were to be purchased at or through it. High grade shoes and cloth might be purchased; shirts and underclothing made to order and fresh flowers delivered. All the products of the Quartermaster Farm were sold, in addition to articles usually dealt in by the ordinary general sales store. Its service extended to several thousand people.

In connection with this undertaking, mention should be made of the Salvage Division, which reclaimed used property and disposed of damaged articles in such a way that it saved the government many millions of marks during the year. The laundry and dry-cleaning plant, operated in conjunction with the Salvage Division, continued to render valuable services throughout the period of occupation.

Transportation.

Organisation.

The transportation activities of the A.F.G., under the Quartermaster, continued to be closely connected with the G-4 control of railways and waterways in occupied territory, the majority of the personnel performing duties with transportation being connected with the Interallied Railway Commission or the Interallied Waterways Commission.[1]

Although it was modified somewhat in 1922 to meet changed conditions resulting from the large reductions in the strength of the A.F.G., the following outline shows the transportation organization, exclusive of the motor branch:

Embarkation
- (1) Preparation of troops for return to United States.
- (2) Base Ports.

Rail and Inland Water
- (1) R. T. O. Service.
- (2) Railway Operation.
- (3) Barge Transportation.
- (4) American Interallied Waterway Commission.
- (5) American Interallied Waterway Commission.

In this Department, as in all others, 1922 was a period of activity and change. Early in the new year the Motor Transport Section became a separate activity reporting direct to the Chief Quartermaster.

Embarkation.

This section continued to assign officers to transports

---

[1]These two commissions are treated in Chapter 3.

and to quarters thereon; to procure trains and equipment necessary to transport personnel and baggage to ports, and to make all arrangements incident to the move with the proper authorities.

All troops were routed from Coblenz to Antwerp by rail and thence to the United States by the Army Transport Service during the first half of the year. These movements included large shipments of baggage and supplies. The last movement toward Antwerp was just prior to the sailing of the "Cantigny" on June 16th, after which all activities along the Coblenz-Antwerp line were discontinued.

Base Ports.

The Port of Antwerp and Port of Bremen are discussed later in sub-chapter G.

Lines of Communication.

Even in the spring of 1922, prior to the receipt of orders for the withdrawal of the A.F.G., a survey had been made of the harbors and facilities at Bremen, Hamburg, Emden and Wilhelmshaven, with a view to transferring the port activities of the A.F.G. from Antwerp to some German port for economical reasons. Work on this plan was naturally discontinued on the receipt of the evacuation orders in March; but after instructions were received for the maintenance of a small force on the Rhine, and after the Army Transport Service between America and Europe was discontinued, the base port of Antwerp was closed on June 20th, and the matter of obtaining the use of a favorable German seaport was revived.

Upon authorization of the War Department, negotiations with German government officials of Berlin were begun in July with the object of effecting an agreement whereby a line of communication should be established with the use of a German port, the operation of which should be at the expense of the German government. These negotiations resulted in what is known as "the Protocol of August 11th". The outstanding feature of this agreement provides that the expense involved in moving United States government personnel and freight between New York and Coblenz, and vice-versa, as well as from intermediate points, be paid direct by the German government. Although the operation along these lines of communications proved entirely satisfactory and the German government officials cooperated fully in carrying out the provisions of the protocol, the necessity for a secondary line of communications not touching unoccupied territory remained.

The contract with the Belgian government providing for a line of communications between that country, with Antwerp as a base port, expired on November 6th, 1922; but negotiations resulted in a new agreement retaining the advantages of the old contract and made possible the use of this secondary line with Antwerp as a base port in case of disorder or threatening trouble in unoccupied Germany.

The advantages of this secondary line were very apparent when, following the order for withdrawal in 1923, the U.S.A.T. "St. Mihiel" called at Antwerp to transport the combatant troops of the A.F.G. to the United States. All other personnel and the surplus supplies not disposed of in Germany were routed through unoccupied territory to Bremen, from which port the evacuation to the United States was effected.

Settlement of Belgian Railway Accounts.

The Belgian government was paid $120,000.00 in cash for supplies and services rendered the American Forces incident to the operation of the line of communication by way of Antwerp prior to June 30th, 1921. The cost of operating over the same line for the period ending June 30th, 1922, was $53,799.15.

Instead of paying this last-named amount in cash, the Chief Quartermaster, A.F.G., made a recommendation that the Belgian government be given credit on account of the notes due the United States. This recommendation was approved by the War Department and upon being submitted to the Belgian government for consideration, was accepted. Therefore, under this arrangement, the Belgian government was given credit on notes due the United States for approximately $54,000.

Rail and Inland Water.

Railway transportation matters continued to be looked after by the R.T.O's. (Railway Transportation Officers), although the number of R.T.O. posts was diminished greatly and the personnel on duty with the remaining ones much reduced. From an organisation which maintained ten R.T.O. posts in occupied Germany, one in France, one in Luxemburg and two in Belgium in the early part of 1922, its activities were so curtailed that there existed R.T.O. posts at Coblenz, Coblenz-Luetzel,

Cologne and Paris only during the last half of the year. In addition to their duties in connection with transportation of personnel, equipment and supplies and rendering assistance to casual army or civilian travelers, the personnel on duty at these posts performed the duties of Military Police in many instances.

The Train Police, organized to supervise the conduct of Allied military personnel on trains in the occupied territory and to enforce the regulations of each Army in its respective area and the instructions of the Interallied Railway Commission concerning the occupancy of military reservations, etc., was continued on a reduced scale. The performance of this duty, which required selected men with considerable tact and patience, in addition to a modicum of linquistic ability, was carried out in a very creditable manner.

Railway Operations.

The 16th Battalion, Railway Engineers, continued to function under the Transportation Officer in its training in operation of a section of railway during the early part of the year, and was prepared to operate trains to maintain communications with the base ports should strikes have necessitated. Training on operating lines had to be discontinued after the evacuations of May and June, in order that personnel from this organization (later designated as Detachment 1st Engineers) might replace Quartermaster troops on duty with the Interallied Railway and Waterways Commissions.[1]

---

[1]See Chapter 3.

Barge Transportation.

Barge transportation between Antwerp and Coblenz which, due to low railway freight rates, had decreased greatly in the latter part of 1921, finally ceased entirely in 1922. The discontinuance of the Port of Antwerp in June excluded altogether the use of this method of transporting supplies and equipment.

Interallied Railway and Waterways Commissions.

These are discussed in Vol. 2, Chapter 14, A.R. in O.G., 1920-21.

Motor Transport Service.

The Motor Transport Service continued to serve the needs of the American Forces and, in addition, did a great deal of repair work for the Graves Registration Service and American Embassies, Legations and military attaches in Europe.

Activity in this branch of transportation was not restricted in proportion to the reduction of strength of the A.F.G. On account of the scarcity of trained enlisted personnel, it became necessary to employ civilian chauffeurs, mechanics and laborers to operate and maintain transportation in a proper state of repair. The employment of these civilians was without expense to the government, as they were paid with funds advanced by a German government. Certain difficulties were encountered in obtaining and retaining expert mechanics in the summer of 1922, on account of the rapid and continued depreciation of the German exchange rate. This difficulty was overcome by certain increases in mark wages which met outside competition and averted further difficulties.

During the period July 1st, 1922 - January 27, 1923, the average daily operation of motor transportation - passenger, truck and motorcycle - was 55. A comparison between the number of vehicles operated on July 1st, 1922, and on January 27th, 1923, shows that there was a diminution from 155 on the former date to 77 on the last-named date.

Officers.

The personnel of the Fourth Division, General Staff, since January 1, 1922, is given below:

		From:	To:
A.C. of S. G-4	Lt. Col. H.H.Sheen, Q.M.C.	Jan.1,1922	Withdrawal
Assistant	Maj. R.B.Patterson, Cav.	Jan.1,1922	June 10,192
"	Maj. Edwin A. Bethel, C. of E.	July 2,1922	Dec.13,192
"	Maj. A.G.Strong,C.A.C.	July 31,1922	Withdrawal
"	Capt. John C. Platt,Jr., Inf.	Jan.1,1922	May 9, 192
"	Capt. I.A.Crump, Ord. Dept.	July 2,1922	Withdrawal

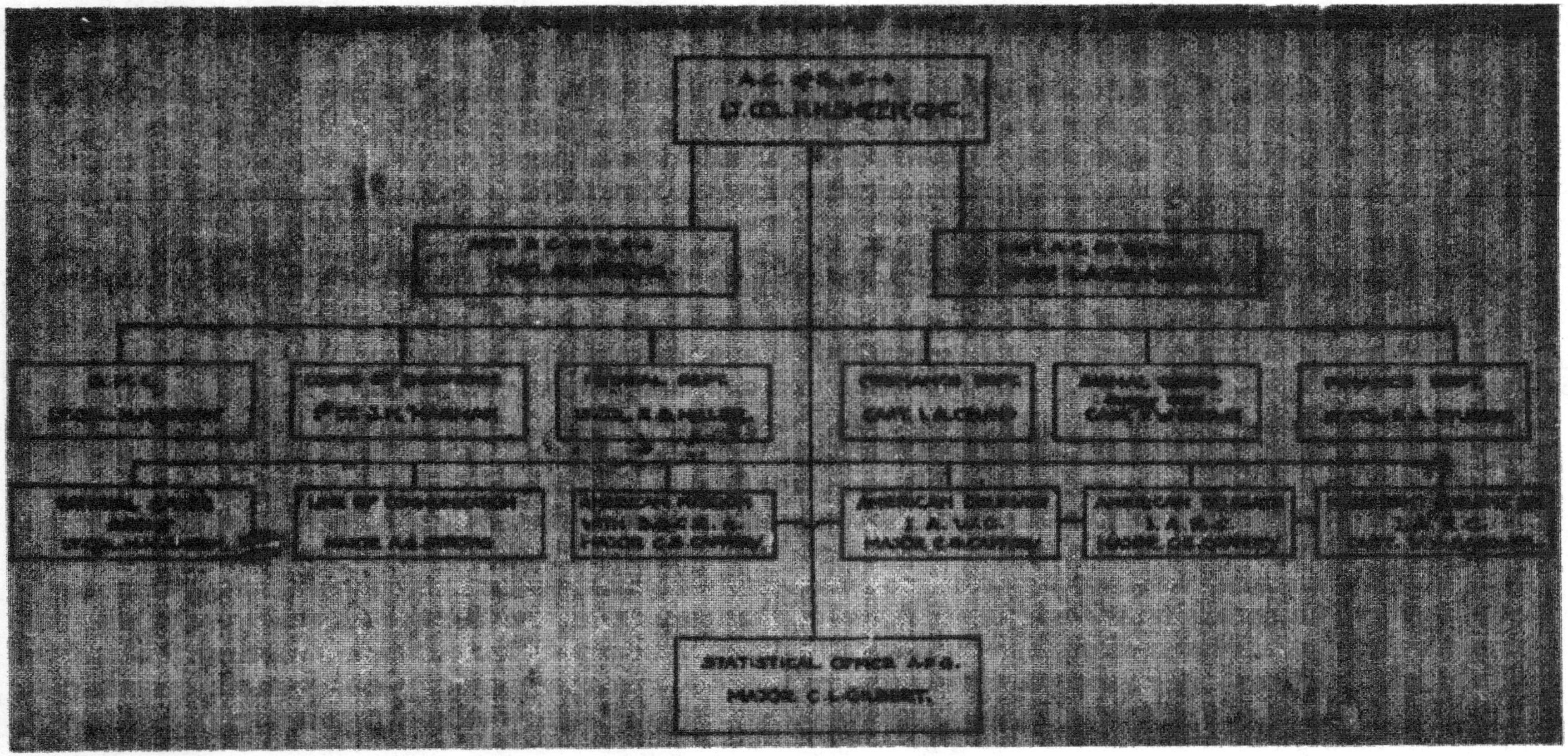
A.C. of S., G-4
LINE OF COMMUNICATION
AMERICAN DELEGATE
STATISTICAL OFFICE A.E.F.
MAJOR C. L. GILBERT.

Pursuant to notes exchanged dated the 5th and 7th July, 1922, and 6th and 8th July, 1922, respectively between duly authorized representatives of the War Department of the Republic of the United States of America and the Foreign Office of the Republic of Germany, duly authorized representatives of the War Department of the Republic of the United States of America and of the Republic of Germany are signing the following agreement as proof of their accord on the principles contained in said notes and as proof of their mutual agreement to the terms contained therein:-

Article 1. A line of communication for the American occupying forces on the Rhine River will be established running from termini in the American Sector through occupied and unoccupied Germany, through the German unoccupied port or ports of Hamburg(Cuxhaven) or Bremen(Bremerhaven) to New York City inclusive, and the Republic of Germany will furnish in kind, or pay for, so many of the services and/or supplies necessary to properly operate such a line of communication thereby effecting for her great economics and making, so far as possible, such obligations and/or expenditures national in character rather than foreign.

Article 2. The water transportation from New York City in America to the German base port or ports, Hamburg(Cushaven) or Bremen(Bremerhaven), in unoccupied Germany, and vice versa, may be accomplished at the option of the War Department of the United States of America by use of vessels of:-

(a) German steamship companies,
(b) Subsidiaries of German steamship companies,
(c) Companies affiliated with German steamship companies,
(d) Companies associated with German steamship companies,
(e) United States Army Transports.

and when none of the above means are available, by any commercial means from any European port. The transportation from the German base port in unoccupied territory to termini in the American Zone, occupied territory, and vice versa, will normally be accomplished by the German State Railways. Besides water and rail transportation in the German base port and along the line of communication, in emergency motor and other transportation may be used. The German government will furnish the necessary rail and other transportation facilities upon request from the American military authorities.

Article 3. (a) Cash payments made by the Republic of Germany for any purpose covered by this protocol which are stated in foreign currencies will be made in marks at the current rate of exchange of the place of payment and on the date of payment, but for accounting purposes between the two parties signatory hereto the conversion rate from marks to dollars or to other currensies will be the rate officially published by the Finance Officer, American Forces in Germany on the date of payment.

All obligations arising from the establishment and operation of the line of communication mentioned in paragraph 1 will be promptly paid by the German Government. The costs to be borne by Germany are calculated from ship's side on pier in New York to and including termini stations in the American occupied territory, and vice versa.

(b) In order to permit the Governments concerned to ascertain the detail costs at any time of the American occupying forces on account of the operation of this line of communication, the American representatives and the German authorities still to be appointed will keep accurate accounts of the expenditures within the following zones:

I. New York to Dover(or other port or ports of call, and vice versa
II. Dover(or other port or ports of call)to German base port, and vice versa.
III. Expenditure in the German base port.

IV. Expenditures from the German base port to the frontier or border of the occupied territory established by the Treaty of Versailles, or Treaty between the United States and Germany, or annexes to them, and vice versa.

V. Expenditures made not along the line of communication but properly pertaining to it.

(c) These expenditures will be divided according to purposes as follows:

I. Obligations which pertain to the American Army of occupation exclusively.

II. All other obligations(see paragraph 9 below).

Article 4. (a) In order to insure the prompt and proper settlement of the German obligations, a German officer with a specially appointed official of the German Government will be established in the German base port or ports(Hamburg or Bremen) and in Coblenz, and will be charged with the payment of all German obligations. Payment will be made on duly executed certificates which have been properly accomplished by the officially designated American representative. These duly accomplished papers must contain an accurate specification and description of the materials shipped; and in case of troop movements the exact strength of the detachment, showing starting point and destination, (see paragraph 6 below), and these documents will be transmitted to the proper German representative by the proper American representative to be later appointed.

(b) The German fiscal representatives will make payment promptly and within five days of the due date of the obligation in cash or other manner suitable to the creditor. The due date is the day on which such obligation is received in the German paying office.

(c) In order to expedite the payment of the obligations arising from water transportation from New York to the German base port or ports(Hamburg or Bremen), and vice versa; to expedite the payment of obligations insurred for water transportation between ports intermediate to New York and German ports, both ways; and to expedite the payment of obligations arising in German base port or ports as well as at intermediate ports, these payments will be made exclusively by a German official yet to be appointed whose place of duty will be at the German base port or ports.

(d) There will be paid at Coblenz by a German fiscal agent yet to be appointed obligations as follows:

I. These arising in operating the line of communication from termini in the American Occupied Area on the Rhine River to the German base port or ports, and vice versa.

II. These obligations from termini in American Occupied Area to and from intermediate ports in Europe on the New York-German port steamship route.

III. These arising not along the line of communication but properly pertaining to it.

(e) If it develops from practical operation of this protocol that obligations must be paid at other places in Germany than those called for herein or that these called for herein should be changed the necessary rearrangement is hereby accorded to effect this, but the other stipulations of Article 4 must be rigorously observed by any paying agencies created.

Article 5. (a) The American Army is entitled to transport at the expense of the German government both ways throughout the line of communication established by Article 1 and 2 of this Protocol, or from termini in the American sector of Occupied Germany overland through any intermediate port of call in Europe or vice versa, the following persons and things:-

I. Personnel of the Armed forces of the United States, personnel appertaining thereto, personnel which by the American laws or regulations are required or permitted to be transported, the dependents of all such personnel, the property of all such persons, and the military stores and supplies of the armed forces of the United States.

II. Upon the written request of the American Representative of the American Department of the Inter-Allied Rhineland High Commission the property of the American Department of the Inter-Allied Rhineland High Commission, the personnel of said Department with their families and their property, which shall be effected in the same manner and under the same conditions as provided herein for the armed forces of the United States of America. The American Military authorities will keep an account of the costs resulting from the transportation of the persons and things of the American Department of the Inter-Allied Rhineland High Commission but in a catagory separate from and in addition to other accounts required elsewhere in this protocol.

(b) It will be necessary for the American Army to use ports of call between New York City and the German base port or ports and to incur obligations, on this account, as well as to incur obligations elsewhere than along the line of communication, but properly pertaining to the operation of said line. It is understood and agreed that costs for travel or shipment by land to and from the termini in American Occupied Area from and to such intermediate ports of call, as well as obligations incurred elsewhere in Europe, but properly pertaining to the costs of operating the said line of communication, will be classed as an obligation under Article 3 above and paid as indicated in Article 4 next above.

(c) Material or property permitted to be transported by this protocol which is shipped in or out of the German Occupied area or shipped through German Territory, through the German base port or ports, will be exempt from incoming or outgoing customs dues in accordance with Ordinance No. 72 as amended of the Inter-Allied High Commission, which ordinance and amendments, insofar as they apply to this protocol, are made and included as integral parts hereof, and will be shipped to and fro as specially exempted customs traffic, using certificates required in the Ordinance and amendments mentioned above.

(d) It so happens that commercial transportation companies provide tickets to travellers which include transportation and subsistence without indicating the value of each element and that dependents and others under American laws and regulations may not be entitled to the subsistence part of the ticket. It also happens that the value of the subsistence part of the ticket varies on the ships of the different steamship companies, therefore to obtain uniformity and simplicity in keeping accounts and to give Germany credit due her, it is agreed that when such commercial tickets are furnished persons who are entitled to transportation but not subsistence the American Military authorities will collect from every such person the value of the subsistence from the class of passage as prescribed by American Army Transport Service Regulations and further that all such collections will constitute credits to Germany(see Article 9, paragraph (c) below). Monthly statements will be prepared and transmitted by the U. S. Army to the proper German officials, showing purpose and amounts of such collections.

Article 6. The transportation of persons and/or things mentioned in paragraph 5 next above between the German base port and the station, destination or terminal in the American occupied zone, and vice versa, will be accomplished on military movement orders or on military waybills, or bills-of-lading similar to the ones now used for Entente transportation in the occupied German territory. The form for the transportation of personnel and material agreed to between the occupying authorities in the occupied German territory and the German State Railways will also apply for military personnel and matériel for railway transportation between the occupied territory and the German base port, and vice versa, changing words "Occupied Territory only" to "Unoccupied Territory only", employing separate serial numbers, the serial number to be preceded by capital letter "U" and followed by the capital letter "S". The French text on the forms for transportation will be omitted.

Article 7. (a) The transportation of American unarmed troops in uniform between the occupied territory and the German port or ports is permitted without special pass and without special permission. Armed guards convoying troop movements or valuable supplies are permitted.

(b) Regarding the presence or circulation of American troops in the city and in and out of the German base port international law and custom will prevail, such as are habitual to the crews of foreign men-of-war when docked in foreign countries, when their crews are ashore, who then adhere to international laws of intercourse.

Article 8. (a) Claims arising from losses or damages along this line of communication for or against Germany, America or third parties, are considered as obligations and will constitute liabilities against the German

Government except in case of "force Majeure", or in a case in which the claiment is to blame; such exceptions would constitute exemption or limitation of the German Government's liability. Since these two latter exceptions may constitute exemptions or limitations of the German Government's liability, these exceptions are made without relinquishing the right of the German Government to make claims against third parties for liabilities created against such third parties. Insofar as such transport or other agreements have been entered into by the American Government, or its representatives, the right accruing to the American Government from these agreements will be transmitted to the German Government for the purpose of their execution. With these exceptions the Republic of Germany guarantees the United States of America against loss by claims emanating from third parties and resulting from the operation of this line of communication.

(b) For the settlement of claims between the Republic of the United States of America and the Republic of Germany arising out of and-or under the application of the terms of this protocol, a mixed Board of Arbitration consisting of three Americans and three German numbers will be established at Coblenz, appointed respectively by the Commanding General, American Forces in Germany and the German Government. The Board of Arbitration will effect its own organization and choose a chairman who will be a seventh member. Its duties shall be to initially consider and investigate all differences which may arise and submit its report, recommendations and findings to higher authority for action necessary and in conformity with the laws of each contracting government.

Article 9. (a) The execution of the above agreement may result in expenditures which according to paragraph 3 above must be paid in the first instance by Germany, but which may not be chargeable to the American Army of Occupation's upkeep. The American Army will therefore keep accurate account of expenditures which will be properly classified according to purpose and zone in proper order and will furnish the designated representatives of the German Republic with certified copies thereof at fiscal intervals to be determined later(see Article 9, paragraph (c) below).

(b) Differences of opinion which may result as to category and classification of all liabilities connected with the operation of the line of communication shall in all cases be referred to the Board of Arbitration for initial consideration and investigation. This board will submit to the respective adminstrative chiefs of the governments concerned their conclusions, recommendations and findings for further administrative action as required by the respective laws of each contracting party.

(c) All payments made by the German Government not properly chargeable to it as a cost of the upkeep of the American Army of Occupation under this protocol shall be placed to the credit of the German Government on account of monies due the Government of the United States of America for the cost of the upkeep of the American Army of Occupation. It is agreed that only persons and materiels mentioned in paragraph 6 above, and who properly come under the title rights created shall be permitted to use the line of communication mentioned in paragraph 1 at the expense of Germany.

Article 10. Should America join in the agreement made by Finance Ministers of other Allied Powers with respect to Germany on March 11, 1922 ("i.e., change the definition of what charges constitute proper costs of Maintenance of the American Army of Occupation"), of the ratification of which the Government of Germany was officially notified, the provisions of such new arrangement will take the place of this agreement insofar as there is a conflict.

Article 11. (a) Should either of the participating Governments feel that they wish to cancel the foregoing agreement for any purpose, such cancellation must be proceded by thirty(30)days' written notice.

(b) All former agreements made by the parties hereto inconsistent with this agreement are hereby revoked. (The "former agreements" mentioned here refer only to any former agreements pertaining to the line of communication as covered in other protocols or addenda negotiated by the individuals signatory hereto).

(c) All obligations incurred on account of the operation of the line of communication since July 15, 1922, under the terms of the cancelled protocols or addenda will be settled under this protocol.

Article 12. The English and the German texts of this agreement shall be equally authoritatiive and shall become effective on the date of signature. Signed, sealed and delivered this the 11th day of August, 1922.

The Authorized Representative of the German Republic.

(signed) Dr. H. Ronde
DR. H. RONDE
Ober-Reglerungarat im Reichsschatz ministerium.

The Authorized Representative of the War Department of the United States of America.

(signed) W. C. Koenig
W. C. KOENIG
Major, C.A.C.(DOL)
U.S. Army.

WITNESSES:

As to signature of the authorized representative of the German Republic.

As to signature of the authorized representative of the War Department of the United States of America

(signed) A. R. Loustalot,
Major, C. A. C.,
U.S. Army.

APPROVAL RECOMMENDED:

(Signed) H.H. Sheen
H.H. SHEEN
A.C. of S., G-4

APPROVED 14 August, 1922.

(Signed) Henry T. Allen,
HENRY T. ALLEN,
Commanding General,
American Forces in Germany.

(WCK/wmb)

Appendix No. 2(f) Headquarters Chapter IV.

AMERICAN FORCES IN GERMANY

Office A.C. of S., G-4

WCK-wmb

AMENDED COPY
(Changes appear in Capital Letters).

Coblenz, 14 October, 1922.

Subject: Letter of Instructions - Line of Communication.

To: Chiefs of General Staff Divisions
Chiefs of Services, G-4.
Officers, L. of C.

1. (a) A Line of Communication for these Forces has been established under authority of Section 2, Joint Resolution of the Congress of the United States approved July 2, 1921, and Article 249 of the Treaty of Versailles and Annexes thereto by a Protocol. A copy of this Protocol is attached hereto for the information and compliance of all concerned.

(b) The A.C. of S., G-4, these Headquarters, is charged with the preparation of plans and policies, and supervision of activities concerning transportation by land and water, including ports of embarkation and the necessary auxiliaries under this Protocol.

2. For the Board of Arbitration required in Article S] subparagraph (b), American members are appointed as follows:

Major A. C. Strong, C.A.C. (DOL)
Captain John Biggar, Q.M.C.
Captain W. H. Sadler, Q.M.C.

3. (a) Major A. C. Strong, C.A.C. (DOL) is appointed as Commander, Line of Communication. He will be an assistant to the A. C. of S., G-4, these Headquarters, and is charged with the carrying out of the policies and plans of the A.C. of S., G-4, wherever the Protocol with Germany is directly affected.

(b) The Fiscal and Records Office will be the office of record for the Commander, Line of Communication.

(c) All correspondence on any subject pertaining to this line of Communication will be transmitted through the Commander, Line of Communication, these Headquarters, or through his staff.

(d) All correspondence from the Commander, Line of Communication, or his staff, to Services or others pertaining to operations along Line of Communication may be direct.

(e) The following abbreviations are authorised:

I.

L. of C.	Line of Communication.
C.L. of C.	Commander, Line of Communication.
C.T.O.	Chief Transportation Officer.
F.&R.O.	Fiscal and Records Office

II. MEANING OF SYMBOLS "AO" and "US" WHEN USED IN CONNECTION WITH L. OF C.	"AO"	"Chargeable to and payable by Germany on account of cost of t American Army of Occupation and non-reimbursible."
	"US"	"Payable in the first instance by Germany but afterwards to be reimbursed by credit on account of monies due the American Government for past costs of the maintenance of the American Arm of Occupation.

4. Captain J. J. McConville, Q.M.C., is appointed as the officer to function at the German base port or ports, as required by Article 4, sub-paragraph (c), of the Protocol, and will be a staff officer of the Commander, Line of Communication, on matters pertaining solely to the protocol.

5. Captain John Biggar, Q.M.C., is appointed as officer to function at Coblenz, as required by Article 4, sub-paragraph (d), sections I, II, III, of the Protocol, and, in addition to his other duties, will be a sta officer of the Commander, Line of Communication, on matters which pertain solely to the Protocol.

6. The following personnel is authorised for the Line of Communicat

(a) at COBLENZ -

Commander, Line of Communication.
Major A.G. Strong, C.A.C.(DOL)

1 stenographer — $90.00 per month, with billet

Chief Transportation Officer, A.F.G.
Captain W. H. Sadler, Q.M.C.

1 interpreter — $25.00 per month, with billet

Fiscal and Records Office.
Captain John Biggar, QMC.
Field Clerk, Q.M. Corps.

1 statistisian	$300.00 per month, with bille
1 clerk	100.00 " " " "
1 stenographer	90.00 " " " "
1 interpreter	Changes as Mark varies - not exceed 20,000 Marks per month

(b) At BREMEN (Port Office)

Captain J. J. McConville, Q.M.C., Port Officer.

1 clerk	$100.00 per month, with bille
1 stenographer	90.00 " " "
1 interpreter	Changes as Mark varies - not t exceed 20,000 Marks per month.

All civilian employees authorised herein will be paid in Marks, as provided in Article 3, subparagraph (a) of the Protocol of 11 August, 1922.

7. (a) As heretofore, the Chief Transportation Officer, A.F.G., will receive and act on all requests for transportation of personnel and material in Europe, and will define the class, port and other conditions under which transportation will issue, and after issuing such authorisation he will transmit one copy to the Fiscal and Records Office, one to the port Officer, one to the Commander, Line of Communication, and two to the applicant, keeping one copy on file in his office. Every copy of approved transportation request, except the one on which transportation is to be issued, will have stamped across its face "ADMINISTRATIVE COPY ONLY - NOT GOOD FOR TRANSPORTATION". In case water transportation is to be used through an intermediate port, an additional copy of such authorization will be sent to the Agent at the Port in question. The transportation will then ensue in accordance with the instructions on approved transportation requests, and will be carried out at the port by the Port Officer or Agent there. In the case of agents at intermediate ports, they will send the original accomplished transportation request to the Port Officer, Nord Deutscher Lloyd Building, Bremen, Germany. The Port Officer at the German base port will make the reports required by War Department regulations to the Chief Transportation Officer, A.F.G., who will then transmit them to the Fiscal and Records Office for final accounting, action and record. In case that water transportation issue at any other ports in Europe than above, the Agent at port will return the accomplished transportation request to Chief Transportation Officer, A.F.G., Coblenz, for visa and check. The C.T.O., will then transmit it to the Port Officer for payment.

(b) In securing transportation under the last two lines of Article 2 of the Protocol quoted as follows: "The German Government will furnish the necessary rail and other transportation facilities upon request from the American military authorities," the following procedure will govern:

RAIL TRANSPORTATION -

In Germany as provided in Article 6 of the Protocol.

In France and Belgium the customary O.D.T's will be issued and rights of Germany as secured in lines 10 - 14, Article 8 (a) of Protocol, will be conserved.

In other countries where the Protocol is involved, by special arrangements between the railways concerned and the Chief Transportation Officer, AFG.

MOTOR TRANSPORTATION--

When needed, requests will be made on nearest German Transportation Representative by the Commander, Line of Communication.

OCEAN TRANSPORTATION -

All requests for ocean transportation will be transmitted through the port Officer to the Agency or Agencies designated by Germany, and they must provide it at minimum rates by any of the means listed in subparagraphs (b)

(c) or (d), Article 2, of the Protocol, using only vessels of American Registry, but if such vessels are not available then by any vessel from any European port. The Chief Transportation Officer, A.F.G., may elect the port and vessel where any choice exists.

(c) For the purpose of supervising transportation activities along the Line of Communication, The Chief Transportation Officer, A.F.G., represents the Commander, Line of Communication, on all transportation matters affecting the Protocol and may issue any necessity orders or instructions in this capacity to properly supervise such transportation matters.

8. All orders issued which involve an obligation along the Line of Communication will be issued by the Adjutant of these forces in the usual way upon request of the A.C. of S., G-4, the Commander, Line of Communication, of any of his staff.

9. (a) Referring now to the system of accounting and payment of obligations along this Line of Communication: The German Government has designated German Fiscal Agents, one stationed in Coblenz and one stationed at the German base port. All obligations properly payable under the Protocol, and all papers pertaining to the procurement or collection of supplies or services along this Line of Communication will bear the symbols "AO" or "US", THE SYMBOL "AO" WHEN APPEARING OR ANY L. OF C. PAPER WILL MEAN "CHARGEABLE TO AND PAYABLE BY GERMANY ON ACCOUNT OF THE COST OF THE AMERICAN ARMY OF OCCUPATION AND NON-REIMBURSIBLE". THE SYMBOL "US" IMMEDIATELY PRECEDING THE PROCUREMENT AUTHORITY AND PURPOSE NUMBER WILL MEAN "PAYABLE IN THE FIRST INSTANCE BY GERMANY BUT AFTERWARDS REIMBURSIBLE BY CREDIT ON ACCOUNT OR MONIES DUE THE AMERICAN GOVERNMENT FOR PAST COSTS OF THE MAINTENANCE OF THE AMERICAN ARMY OF OCCUPATION". In carrying out the payment of these obligations all military personnel traveling or shipments made to or from these Forces will be forwarded with transportation charges, C.O.D. from New York, and prepaid to New York, including advance payments for tips and steamer chair on steamer, except that no advance payments will be made to officers or others traveling under the symbol "US". The Port Officer will require each person to whom payment is made in advance for tips and for steamer chair to make a certificate that this money will be used for purposes for which advanced, and that on arrival at destination in the United States such person will execute and return a certificate that this money was used in the manner and for the purpose provided. The Post Officer will then transmit this certificate through the Chief Transportation Officer, A.F.G., for final file in the Fiscal and Records Office, these Headquarters. If any unused balances of such advances remain on hand on arrival in the United States, they will be turned in by person concerned to the nearest Finance Officer for credit to "Miscellaneous Receipts", Fiscal Year involved and such Finance Officer should so advise the Chief Transportation Officer, A.F.G., Coblenz, Germany, who will transmit the advice to the officer who approved the bill for payment.

(b) The War Department has been requested to transmit to the American Port Officer, Nord Deutscher Lloyd Building, Bremen, Germany, all obligations involved under the Protocol for payment and that the symbols given above be written on such documents.

10. (a) IT WILL BE NECESSARY IN KEEPING THE ACCOUNTS OF THIS NEW LI

OF COMMUNICATION TO KEEP ALL ACCOUNTS BY ZONES AND CLASSIFICATION, AND TO ENTER ON ALL OBLIGATIONS, EXCEPT "AC" ACCOUNTS AND THOSE PERTAINING TO THE MEMBERS OF THE AMERICAN DEPARTMENT, INTERALLIED RHINELAND HIGH COMMISSION PROCUREMENT AUTHORITIES AND PURPOSE NUMBERS. EACH "AO" OBLIGATION WILL HAVE ENTERED THEREON THAT IT IS CHARGEABLE AS AN "AO" ACCOUNT UNDER AUTHORITY OF WAR DEPARTMENT CABLEGRAMS W-1101 AND W-1102, OCTOBER, 1922. Germany will make all initeial payments, but will be entitled to credit for the "US" accounts only. The methods of recording and reporting obligations of and expenditures by the Republic of Germany under the symbols "AO" and "US", and for the American Department, Inter-Allied Rhineland High Commission, are prescribed in subparagraphs (b), (c) and (d) below. The proper supporting documents (Travel order, movement order, transportation request, authority for reimbursement of actual expenses of officers, etc., commutation of rations for enlisted men, etc) for all accounts passed to the Republic of Germany for payment will be attached to and retained with the retained copy of voucher in the office of the Fiscal and Records Officer, Line of Communication.

(b) All "AO" ACCOUNTS (OBLIGATIONS WHICH ARE A PROPER CHARGE AGAINST GERMANY) WILL BEAR THE SYMBOL "AO", AND A SEPARATE ACCOUNT WILL BE KEPT OF ITEMS UNDER THIS SYMBOL. The Fiscal and Records Office will report such expenditures monthly, as of the last day of the month, to the Finance Officer, A.F.G., through the Statistical Officer, A.F.G. for his information as services furnished in kind by the Republic of Germany.

(c) All "US" accounts(obligation for which the Republic of Germany is entitled to credit) will be stated on forms which have been approved by the Finance Officer, A.F.G., and the A.G. of S., G-4, for use in carrying out the terms of the Protocol. These forms will carry the necessary data to make them acceptable to Finance Officers, United States Army, so that proper credit may be given the Republic of Germany on the unpaid balance of the cost of the American Army of Occupation, and so that Charge may be made against the appropriations properly chargeable with the expenditures. All "US" accounts paid by Germany during the month will be combined on voucher form 330 by the Fiscal and Records Officer, which supported by the copies of each individual form and its supporting documents, will be forwarded as of the last day of the month by the Chief Quartermaster, A.F.G., to the Quartermaster General of the Army, with the request that it be handled in the manner outlined in the beginning of this subparagraph. A copy of this form 330 will be furnished the Finance Officer, A.F.G., through the Statistical Officer, A.F.G., for his information and action.

(d) Statements of accounts incurred for the American Department Inter-Allied Rhineland High Commission, will be prepared monthly by the Fiscal and Records Officer, as of the last day of the month, and will be forwarded in duplicate to the American Representative, Inter-Allied Rhineland High Commission.

(e) RESCINDED.

11. In the use of the United States Lines from Bremen the intermediate ports of Southampton and Cherbourg will be available both ways; in the use of the United American Lines from Hamburg the ports of Southampton and Cherbourg will be available on the west-bound voyage and Plymouth and Boulogne on the east-bound. This office will later appoint Agents at these ports to

care for our interests. These gents will be instructed to send their bills to the American Port Officer, Nord Deuscher Lloyd Building, Bremen, Germany, for payment and necessary administrative action.

12. In order to properly coordinate all transportation activities in connection with the Line of Communication covered in the Protocol, there is now centralized at these Headquarters the transportation facilities and personnel to effect movement of all military personnel and supplies in Europe, as well as others entitled under War Department rules and regulations to such transportation. Commercial facilities to be used for ocean transportation will be considered in a category similar to the United States Army Transport Services and the rules and regulations pertaining thereto will be applied wherever possible. Whenever requests are received from military personnel in Europe they will be acted on accordingly.

13. THE USE OF THE SYMBOLS "AO" AND "US" PRESCRIBED HEREIN SHOULD PRESENT NO DIFFICULTY TO FINANCE OFFICERS IN THE UNITED STATES; if orders are issued by the War Department instructing such Finance Officers to make no payment of transportation allowances or charges on any such orders other than from ship's side, New York, or other port used(except tips and steamer chair costs under symbol "US") to destination in the United States, that is to say, the symbols "AO" and "US" on an order should mean to Finance Officers in the United States that all transportation charges have been paid from these Forces up to and including ship's side, New York(except tips and steamer chair costs westbound under symbol "US") and shipments of material or personnel from New York to these Forces will come C.O.D.: European port.

14. The close coordianation and cooperation of the Commander, Line of Communication, Officer in Charge at Ports, Officer in Charge of Fiscal and Records Office, and the Chief Transportation Officer, A.F.G., will be necessary to carry out the details and the spirit of the Protocol, and at the same time see that the operations are in consonance with the requirements of these Forces.

15. Attached hereto marked Exhibit "B" is a chart showing the organization of these Forces, as well as the channels of correspondence and the relation between initiation of policy, command and operations, which bring within American law and regulation an organization to function under the Protocol.

16. LETTER OF INSTRUCTIONS - LINE OF COMMUNICATION, R.A.F.G., COBLENZ, 30 AUGUST, 1922, IS RESCINDED.

(Signed) H. H. Sheen,
A. C. of S., G-4.

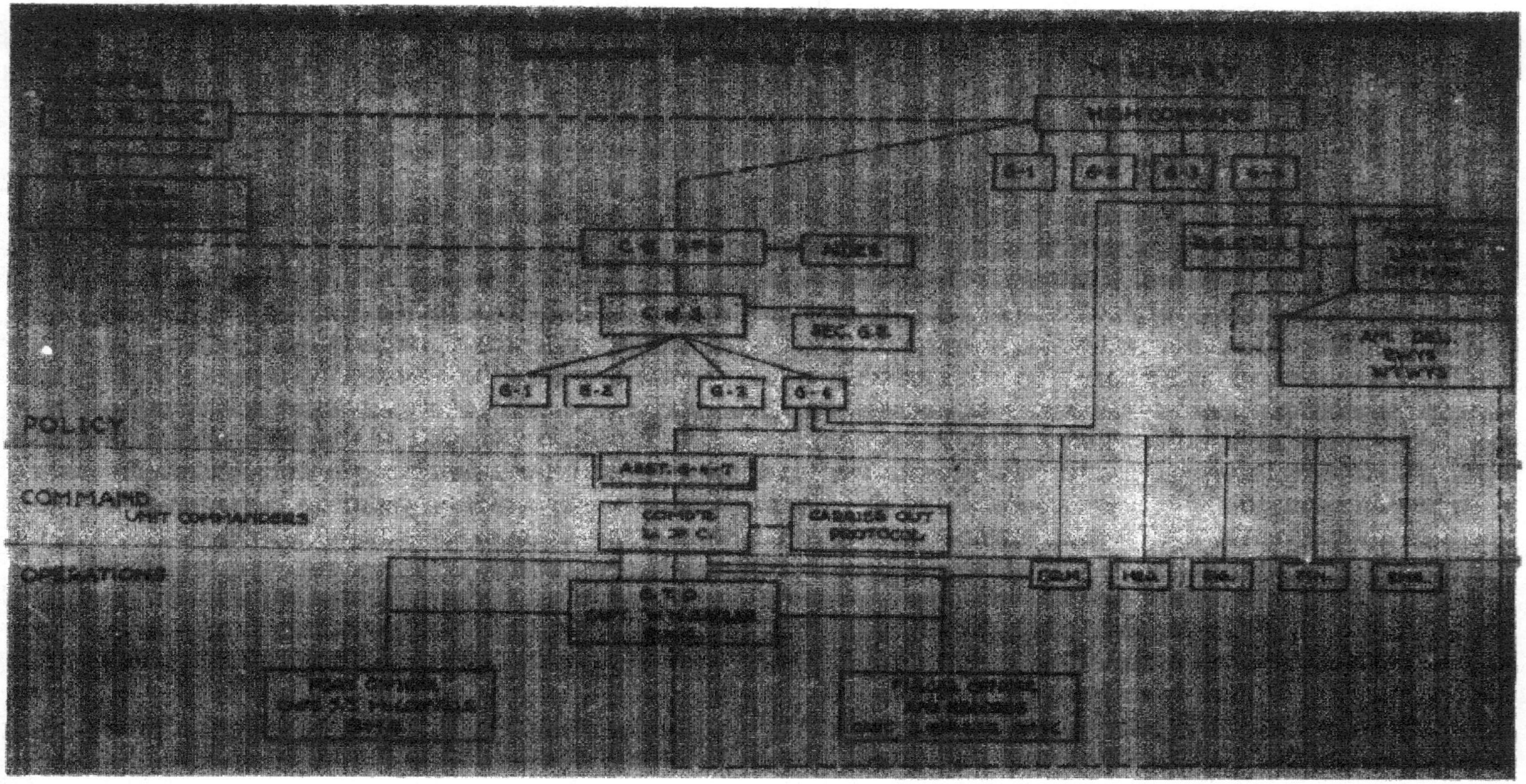
POLICY
COMMAND
UNIT COMMANDERS
OPERATIONS
G-1
G-2
G-3
G-4
SEC. G.S.
CARRIES OUT PROTOCOL

LINE OF COMMUNICATION.

PASSENGER TRANSPORTATION, L. of C., July 21 to Dec. 31, 1922.

WEST BOUND
(East Bound - None).

FROM:	BREMEN			HAMBURG			CHERBOURG			SOUTHAMPTON			TOTAL OF			TOTAL PASSENGERS ALL CLASSES
Class:	1	2	3	1	2	3	1	2	3	1	2	3	1	2	3	
"O" Accounts																
Officers	19	-	-	6	-	-	-	-	-	1	-	-	26	-	-	26
Dependents	15	-	-	7	-	-	-	-	-	-	-	-	22	-	-	22
Enl. Men	-	3	58	-	-	21	-	-	-	-	-	-	-	3	79	82
Civilians	1	-	-	-	-	-	-	-	-	-	-	-	1	-	-	1
Total	35	3	58	13	-	21	none			1	-	-	49	3	79	131
"S" Accounts																
Officers	9	-	-	7	-	-	11	-	-	2	-	-	29	-	-	29
Dependents	6	-	-	12	-	-	5	-	-	1	-	-	24	-	-	24
Enl. Men	-	-	1	-	-	-	-	-	-	-	-	-	-	-	1	1
Civilians	4	-	-	-	-	-	10	-	-	-	-	-	14	-	-	14
Total	19	-	1	19	-	-	26	-	-	3	-	-	67	-	1	68

TOTAL PASSENGERS -	Bremen	Hamburg	Cherbourg	Southampton	All Ports
"O" Accounts	96	34	none	1	131
"S" "	20	19	26	3	68
	116	53	26	4	199
Other passengers - all ports - no expense to Government					34
Total handled					233

SUMMARY BY STEAMSHIP.

Steamship	Passengers.
America	72
George Washington	36
Resolute	30
President Roosevelt	16
Reliance	12
Manchuria	9
President Harding	8
Mount Clay	4
Mongolia	3
Sierra Nevada	3
President Arthur	2

President Fillmore	2
President Adams	1
President Van Buren	1
	199

NOTE: "Other passengers" include dependents, not entitled to transportation at Government expense, of passengers entitled to transportation, and for whom arrangements for sailing were made through this office.

LINE OF COMMUNICATION.

REPORT OF FREIGHT HANDLED.

July 21 to December 31, '22.

EAST AND WEST BOUND.

	Kilo Tons (2220 lbs)		TOTAL Kilo Tons (2220 Lbs)
EAST BOUND			
New York to Coblenz via Bremen -			
Quartermaster Supplies	255		
Motor Transport Supplies	6		261.
WEST BOUND			
Coblenz to New York via Bremen -			
Household goods and personal effects of officers			61.
		TOTAL	322.
Horses - Coblenz to New York			2

LINE OF COMMUNICATION.

COSTS PAID BY GERMANY UNDER THE PROTOCOL
July 15 to December 31, 1922.

1922		"AO" ACCOUNTS	"US" ACCOUNTS	TOTAL.
July		-	-	-
August		-	-	-
September		$4,941.82	$2,278.00	$7,219.82
October		9,830.97	4,281.31	14,112.28
November		5,469.63	4,424.94	9,894.57
December		3,235.93	3,206.48	6,442.41
	TOTALS	$23,478.35	$14,190.73	$37,669.08

Appendix No. 6 (F)

HEADQUARTERS
AMERICAN FORCES IN GERMANY
COBLENZ

HAUPTQUARTIER
AMERIKANISCHER STREITKRAEFTE IN DEUTSCHLAND
COBLENZ

1. The bearer, ______________________________
1. Inhaber dieses,
a member of the American Forces in Germany, is
Angehoeriger der Amerikanischen Streitkraefte in Deutschland,
under military orders to proceed to NEW YORK
begibt sich unter Heeresbefehl nach NEW YORK
on board the ______________________________
an Bord des Dampfers
which is scheduled to depart from ______________________
welcher planmaessig am
on or about ______________________________________
von abfaehrt.

Place photograph here
Lichbild hier aufkleben.

By order of the Commanding General:
Auf Befehl des kommandierenden Generals:

(S) S. G. Talbott
S. G. TALBOTT
Adjutant.

Countersigned:

H. HOSSFELD
Acting A.C. of S., G-2, A.F.G.

OFFICE OF THE REPRESENTATIVE OF THE
FOREIGN OFFICE ATTACHED TO THE GERMAN
COMMISSIONER FOR THE OCCUPIED ZONE.
DER VERTRETER DES AUSWAERTIGEN AMTS BEIM
REICHSKOMMISSAR FUER DIE BESETZTEN
RHEINISCHEN GEBIETE.

1. All German authorities are requested not to interfere with
Saemtliche deutschen Behoerden werden ersucht, der Reise
the travel of this officer (soldier), whose picture appears
dieses Offiziers (Soldaten), dessen Lichtbild sich in der
in the upper left hand corner of this certificate of
oberen linken Ecke dieses Ausweises befindet, keine
identification.
Hindernisse zu bereiten.

INSTRUCTIONS AND INFORMATION.
ERLAEUTERUNGEN.

1. An officer or soldier who is under orders to return to the
1. Offiziere oder Soldaten, die gemaess Heeresbefehl nach den
UNITED STATES must report to the office of G-2 and submit this form
VEREINIGTEN STAATEN zurueckkehren, muessen sich an das Buero G-2 wenden
with a suitable photograph in order that the necessary vises may be
und dieses Formular mit einer geeigneten Photographie vorlegen, damit
obtained authorizing travel in unoccupied German Territory.
die fuer die Reise durch das unbesetzte Gebiet noetigen Sichtvermerke
beschafft werden koennen.

2. This certificate of identification must be constantly carried
2. Dieser Ausweis muss von dem Inhaber im unbesetzten Gebiet
by the person to whom issued when in unoccupied German Territory, and
stets bei sich getragen und auf Verlangen den deutschen Beamten
shown upon request to German officials.
vorgezeigt werden.

THIS SPACE FOR VISES
RAUM FUER SICHTVERMERKE.

- FREI -
gemaess besonderem Abkommen.

Appendix No. 7(f)

[Certificate issued by French Army. Original in French, English and German. English text only reproduced here.]

Army Stamp.

Chapter IV.

No. 1833

I, the undersigned .... certify that the goods enumerated below

dispatched by .... on 19

to (Consignee) .... are Army property

(or) are the property of a member of the family -- of a member of the Interallied Rhineland High Commission; or of a

member of the family of a member of the Armies of Occupation (2), and have actually been used by him.

Description of goods	Amount	Packing	Special marks

At 192

Signature and official stamp

(English text).

Draft of agreement covering the conditions of utilization of the Belgian Railways by the American Army,

-----oOo-----

ARTICLE 1. This agreement states the conditions under which the Government of the United States of American may transport over the normal gauge roads of the Belgian Railways, troops, supplies, arms, ammunition, materials and other impediments, as well as bodies belonging to the military forces which the United States of America may retain in Europe.

ARTICLE 2. As the presence in Europe of military contingents of the United States of America may coincide with a mobilization of the Belgian Army, this agreement will contain two chapters in which the military transportation of the American Army over the Belgian lines will be considered successively with the suppositions that the Kingdom of Belgium be at Peace and at War.

CHAPTER I.

MILITARY TRANSPORTATION OF THE AMERICAN ARMY IN PEACE TIME.

ARTICLE 3. The following are considered as Military shipments for the American Army: Transportation of troops, animals, arms, ammunition, equipment, supplies, and materials belonging to that Army.

The following are considered as supplies for the American Army: Foods, Subsistence stores, and Forage belonging to the Quartermaster Corps of that Army and intended for the use of same. These shipments will be made on charge accounts for the American Government upon presentation of an Ordre de Transport, such as those now being used, at the station of departure in Belgium or at the station of entry into Belgium of the shipments.

These Ordres de Transport will be settled monthly.

ARTICLE 4. The Belgian Railways will furnish the necessary rolling stock for American Military shipments originating in Belgium, except when the American Army desires to make shipments, in all or in part, by means of rolling stock belonging to them.

The movement of empty rolling stock in expectation of American Military shipments will not lead to the collection of charges if the American Army requests the stock without specifying the point of origin, indicating simply the number of cars and their type, determined in general terms.

On the contrary, any movement of a determined car, or of a group, of determined cars, designated by the American Military authorities, between two determined points and designated by that authority, must lead to the issue of an Ordre de Transport.

Consequently movements of empty rolling stock, as per designations below, must be covered by an O.D.T.:

(a) Trains especially made up, in view of the arrival of American Military Transports, and held in storage at the request of the American Military authorities, whether they are going to be loaded, whether they return to the yards after embarkation, whether they change trackage by order of the American authorities;

(b) Cars especially equipped for the needs of the American military authorities or belonging to the American military, and by this fact rendered improper for commercial shipments;

(c) Cars of a definite type, designated in precise terms by the American military authorities and intended for uses of a military nature, with or without special equipment.

ARTICLE 5. Rolling stock stored on the Belgian Railways at the request of the American military authorities will lead to the following charges:

(a) For cars belonging to the American Army: A sum of 50 francs per car per year will be paid by the American Army for the use of storage tracks in Belgium. The average number of cars in storage on the Belgian lines will be agreed upon by the American Army and the Belgian Railways.

(b) Cars not belonging to the American Army and stored at the request of the American military authorities will be considered as being rented by the American Army under the following conditions:

#1. Freight and guards vans of freight trains at the rate of 3 francs per axle per day.

#2. Passenger cars and baggage vans of passenger trains at the rate of 7 francs per axle per day. These charges for rent are for the days of storage only. The days of movement of cars, loaded or empty, which are covered by O.D.T.'s will be deducted from the period of time for which rent is charged. The charges for use of storage lines and the costs of hire of material, are settled monthly the same as O,D,T,'s.

ARTICLE 6. Military shipments or shipments of subsistence stores belonging to the American Army, specified in Article 3, save ammunition which is handled in Article 8, shipments of bodies handled in Article 9, and the movement of empty cars specified in (a), (b), and (c) of Article 4, will be made over the Belgian lines, under the following conditions:

#1. SPECIAL TRAINS:

(a) For special trains of 300 T. (or less) of freight hauled, excluding the service car, and composed of a maximum of 19 cars or 38 axles, the lump charge applicable per kilometer will be equal to the minimum required by the Belgian rates for the organization of a special train (at present 25 francs).

(b) For special trains of more than 300 T. with a maximum of 800 T. of freight hauled, the service car excluded, and composed of a maximum of 50 cars or 100 axles, the lump charge applicable will be as per letter (a) plus 50 per cent (at present 37.50 francs).

The American Army will furnish an O.D.T., and a Lettre de Voiture, such as is used by the Interior Service, for the complete train. The supply of rolling stock and the dispatch of a special train must be requested at least 48 hours in advance.

#2. Shipments by completely or partly loaded cars will be charged for at the normal commercial rates with a reduction of 20 per cent. The American Army will furnish and O.D.T. and a Lettre de Voiture, such as is being used by the Interior Service, for each shipment.

#3. The movement of special cars carrying American military personnel and especially reserved for the American Army will be charged for at the rate of 5 francs per car per kilometre in case it is a car belonging to the Administration of Belgian Railways or foreign lines, and 4 francs per car per kilometre in case the car belongs to the American Army. The American Army will furnish an O.D.T. for each car for each trip.

#4. Transportation of American Military personnel travelling on charge account for the service, along or in small detachments, in public service trains, will, upon presentation of an O.D.T., benefit of a 20 per cent reduction in the price for ordinary transportation of travelers and baggage, provided the soldiers take 3rd class seats, the non-commissioned officers 3rd or 2nd class, and officers any class.

#5. American Military personnel traveling in uniform, alone and at their own expense, will obtain a reduction of 20 per cent upon presentation at the ticket window at stations, of the certificate of which a sample is attached. This applies to: Soldiers traveling 3rd class only; Non-commissioned officers in 3rd and 2nd class; Officers in any class.

#6. Movements of empty cars, which are covered by an O.D.T., will, on production of a Lettre de Voiture, such as is used by the Interior Service, be charged for under the provisions of Tariff #3, "Cars Rolling on Axles"; these prices being fixed for 2 axle cars and increased by 50% per additional axle.

ARTICLE 7. In case of reduction or increase in tariff 3 for merchandise over the Belgian Lines, the charges provided for in Article 5, letters (a) and (b) and Article 6, #3, above, will be modified in the same proportion.

ARTICLE 8. Shipments of powder, explosives and ammunition may be made under the following conditions, on O.D.T. accompanied by Lettre de Voiture of the Interior Service:

(a) Shipments of 10,000 kilos and less, conditioned in conformity with regulations, may be made in ordinary special freight trains of the American Army and in ordinary freight trains of the commercial service. In this latter case, they will be subject to the regular rates of the commercial tariffs.

(b) Shipments of more than 10,000 kilos, with a maximum of 100 T., must be made by special trains which contain no other shipments and which would be charged for as per article 6, #1 (a) - special trains of 300 tons and less.

ARTICLE 9. Shipments of bodies of American military personnel will be made under the supposition that the cars are loaded with 10,000 kilos, giving no consideration to the actual weight of the bodies loaded, and the rate to apply will be Tariff 3, first class B, without reduction. These shipments will be covered by an O.D.T. and a Lettre de Voiture, same as is used by the Interior Service.

ARTICLE 10. All American Army shipments and movements by rail which necessitate a special train or the adding of a special car to ordinary trains, as well as shipments of powder and explosives, will be requested by the American military authorities from the Belgian military Delegation in the office of the Minister of Railways.

ARTICLE 11. Transportation of mails dispatched by the American Army authorities, addressed to the American Forces in Germany, or vice versa, may be made, not only in the special trains of the American Army, but also in freight cars or in cars contained in other trains. In special trains the charge for transportation will be included in the lump price referred to in Art. 6, #1; in cars belonging to other trains the commercial freight rate will be charged, Tariff 3, first class, less 20 per cent, or Tariff "express" less 20 per cent, depending whether the shipment is made by freight trains or by passenger train, it being understood that the C.D.T. 's, and the Lettres de Voiture, such as that used by the Interior Service, will be marked distinctly so as to indicate the nature of these shipments.

ARTICLE 12. The use of telephonic and telegraphic installations of the Kingdom of Belgium by the American Army will be charged for at normal rates and settled monthly.

CHAPTER II.

AMERICAN MILITARY SHIPMENTS IN WAR TIME.

ARTICLE 13. As mobilization of the Belgian Army would result in changes in the departmental control of the Belgian railways, in part or throughout the entire national territory, the status of the American Army shipments over the Belgian railways in time of War would have to be taken up with the Department of National Defense.

ARTICLE 14. This agreement will take effect on November 7, 1922, at 12 o'clock midnight. It may be discontinued at any time by either of the parties thereto, upon a 3 months' notice in advance to that effect.

Made in duplicate at Brussels,
on September 11, 1922.

For the Government of the Kingdom
of Belgium:
For the Minister of Railways, Marine,
Posts and Telegraphs:

(signed) M. HORN.
M. HORN.

(signed) BAUDOUX.

For the Government of the United
States of America:

(signed) H.H. SHEEN.
H.H. SHEEN.
Lt. Colonel, Q.M.C.,
A.C. of S.,G-4.

APPROVED:

HENRY T. ALLEN,
Commanding General.

SUB-CHAPTER G.

BASE PORTS OF ANTWERP AND BREMEN.

Port of Antwerp.

The base port of the A.F.G. during the first half of the year 1922 was Antwerp, Belgium, and through it passed all personnel and supplies from the United States to the American Forces and vice-versa. The organization of the base port underwent no change during this period and it continued to be under the Army Transport Service, reporting direct to the Chief of Staff, A.F.G.

The business done at this port, both as regards cargo and personnel, was particularly heavy on account of the reduction and partial evacuation of the troops on the Rhine. During the first six months of 1922, eleven transports passed through the port. Between April 3rd and June 16th, seven transports, with 4,907 passengers and 35,751 tons of cargo, arrived at and departed from it. Four days after the sailing of the last transport on June 16th, the port of Antwerp was officially closed. Between June 20th and 30th the final settlement was effected and all of the personnel moved to Coblenz for station. The closing of this port necessitated the release of all cold-storage facilities, docks, offices and other quarters; the termination of all contracts and the temporary discontinuance of the Coblenz-Antwerp line of communication.

As an indication of the pleasant relations which always existed between the American troops and the Belgian authorities of Antwerp, it may be mentioned that the Mayor of that city tendered a public reception to the Port Commander and his st at which time he expressed his regret, and that of the people of Antwerp, on the suspension of the American activities there.

Officers.

The following officers were on duty at the port of Antwerp during its existence in 1922:

Koenig, W.C.	Maj., C.A.C.	Port Commander. Port Transportation Officer.
Doane, Guy H.	Maj. F.D.	Port Finance Officer.
Shipp, W.E.	Maj., Cav.	A.P.M. R.R.& C. Officer.
McClure, C.L.	Capt., F.A.	Adjutant; Port Signal Officer.
Littell, A.L.	Capt., Q.M.C.	Asst. to Port Commander. Property Officer.
Biggar, John	Capt., Q.M.C.	Port Quartermaster.
Shipp, F.B.	Capt., Cav.	Embarkation & Landing Officer; Passenger Traffic Officer. P.E.S. Officer.
Graham, O.L.	Capt., M.C.	Port Surgeon.
Hand, H.V.	Capt., Inf.	Headquarters Commandant; Detachment Commander; Charge Post School
Orsinger, R.	Capt., Inf.	Executive Asst. to Port Quartermaster (In Charge A.T.S. Affairs).
Singer, Paul L.	Capt.,Cav.	Asst. to Port Q.M. (R.T.O. Baggage Officer, M.T.O.); Prison Officer; Assistant to Headquarters Commandant.
Whippler, F.R.	1st Lt.,Q.M.C.	Assistant to Port Q.M. (In charge Allotments and Apportionments, and Assistant to E. & L. O.)
Robinson, V.L.	1st Lt., Q.M.C.	Assistant to Port Q.M. (In charge Sales Commissary and Assistant to R.T.O. and Baggage Officer).
McConville, Jos.J.	Capt., Q.M.C.	Port Officer, American Graves Registration Service.

Port of Bremen.

In order to carry out the functions of the line of communication as provided in the protocol, already mentioned, the post of Port Officer, Bremen, was established on August 11, 1922. This officer served in the dual capacity of port representative of the Transportation Officer and as fiscal representative on the staff of the Commander, Line of Communications. All reservations for personnel and arrangements for cargo space required for the A.F.G., Graves Registration Service, and military attaches in Europe passed through this office. In many instances it also attended to reservations for relatives of military personnel who were not transported at government expense. Since War Department instructions required that transportation be furnished on ships of American registry, practically all transportation was furnished on ships of the United States Lines from Bremen. In some instances, when it was impossible to secure reservations or space on this line, ships of the American Line or the United American Lines were employed.

In addition, the Port Officer was responsible for the auditing and passing of all bills to the representative of the German government for payment. The system followed provided that all bills be submitted by the steamship agencies to the Port Officer immediately after these agencies had rendered service to the United States government. The bills were then audited and approved by the Port Officer and passed to the German fiscal agents for payment. All shipments and movements through Bremen were made without expense to the United States government. Due to the close cooperation maintained by the Port Officer, steamship officials and the German authorities, all bills were usually paid within three days after presentation.

Capt. J.J. McConville, Q.M.C., who was appointed Post Officer, Bremen, at the time that office was established, served as such until the withdrawal of the forces.

SUB-CHAPTER H.

MILITARY CONSTRUCTION AND MILITARY RAILWAY OPERATIONS.

Military Construction.

The period covered by this narrative was one of contraction rather than expansion. Consequently, very little military construction was carried out in these forces. Construction activities, which had been transferred from the Engineer Corps to the Quartermaster Corps in 1921, were confined chiefly to the repair and maintenance of German property and buildings used by the American forces. The Corps continued to order and supervise the repair of roads in this area as required by the Army.

During the latter days of the A.F.G., construction activities were concentrated upon the packing and crating of files, baggage and other such supplies which were being returned to the United States pursuant to the evacuation orders. The material and labor utilised in construction work was procured from the German government without expense to the United States.

Military Railway Operations.

After the changes incident to the reduction in the strength of the A.F.G. had been effected early in 1922, Company "A", 16th Engineer Battalion (Railway), with the personnel comprising the former battalion continued training and operations at Kreuzberg and Kottenheim. It operated six trains daily, excepting Saturdays, Sundays and holidays, at each station.

In addition to their purely military training, such as infantry drill, interior guard, physical drills, schools, target practice, etc., the technical training of this railway unit comprised in general the following:

(a) Observation of methods of operation of German railways;
(b) Study of station and yard facilities in American and adjacent areas;
(c) Procurement of data for use in military operation of railways;
(d) Operation of switch engines in various yards;
(e) Operation of school trains;

(f) Schools of instruction in American and German operating methods.

After the consolidation of all the personnel at Kreuzberg, the practical training of the railway unit consisted in the operation of three trains daily from Kreuzberg to Junkerath, a distance of thirty-five miles. The same methods used on a railway division in the United States were followed and all the personnel worked under the appropriate railway title. The necessary repairs to locomotives and rolling stock were made by the personnel in the round-house at Kreuzberg.

Schools in railway training were continued to prepare personnel to perform efficiently the duties of driver, fireman, conductor and brakeman. These courses covered a complete course in both German and American methods of railway operation. Later, this personnel successfully performed the various duties incident to conducting troop trains and to consignment of supplies to and from the port of Antwerp.

The constant reductions in the personnel brought about a cessation of railway operation, and the Railway Engineers were returned to Coblenz for station in April. Some of the personnel was used in the performance of duties with the Interallied Railway Commission, consisting of R.T.O. activities and control post duty, such as checking trains entering and leaving the occupied territories in connection with imports and exports. Other personnel were assigned to duty with the Interallied waterways Commission, which controlled and regulated river traffic. Duty with these two Commissions scattered the personnel to towns between Coblenz and Wiesbaden in the occupied area and necessitated stationing some of them at Paris, Antwerp and Brussels. On October 31st, in compliance with instructions of the War Department, the railway organisation was placed on the inactive list and all the personnel were absorbed by the Detachment, 1st Engineers, A.F.G., although they continued at their duties with the Railway and Waterways Commissions and their varied duties in connection with the Office of the Engineer, A.F.G.

The following officers of the Engineer Corps were engaged in military railway operations in addition to their other duties during 1922:

Major Edwin A. Bethel, Engineer, A.F.G.
1st Lieutenant John M. Harman, Assistant.

SUB-CHAPTER I.

MEDICAL DEPARTMENT.

General Activities.

The activities of the Medical Department of the American Forces in Germany during this period were largely routine, no new policies[1] having been necessitated in view of the gradual reduction of the forces.

On January 1st, 1922, the Medical Department consisted of 48 medical officers, 8 dental officers, 3 veterinary officers, 3 medical administrative officers, 2 warrant officers, 42 Army nurses and 689 enlisted men. Thereafter, the reduction was progressive until June 19th, after which date there were left in the A.F.G.: 6 medical officers, 1 dental surgeon, 1 veterinary surgeon, 2 Army nurses and 33 enlisted men. Additions were made to the Nurse Corps by the arrival on July 15th of two members of the Corps from the United States, and one medical officer was gained on November 6th by the assignment of an officer who had then returned from post-graduate work at Vienna. On November 30th the only dental surgeon and the only veterinary surgeon on duty with the forces left for the United States to be discharged. With these changes, the Medical Department on December 31st, 1922, consisted of 7 medical officers, 4 Army nurses and 30 enlisted men.

---

[1]See A.R. in O.G., 1920-21, Chapter 26.

On January 15th, 1922, the Medical Department Training Centre at Neuwied, which had been maintained since the previous October as the Provisional Medical Battalion, was definitely closed and the entire organisation moved into the Station Hospital. The personnel of Hospital Company 56 was utilised for Station Hospital duties, Hospital Comapny 58 was demobilised on February 28th, and Ambulance Company 57 continued its work for the purpose of maintaining, operating and repairing the necessary transportation for the station Hospital and ambulance service to outlying commands.

Hospital Agreement with French.

In May, 1922, orders were received from the War Department recalling the A.F.G., the movement to be completed by June 30th, with the exception of a few officers and men left as a "clean-up squad". The result of this order was that large French garrisons, totalling some 4,000 men, were sent into the American area. As our forces were thought to be approaching the vanishing point and the French were gradually increasing in numbers, the Station Hospital was turned over to the latter on June 3rd, 1922. In the meantime a small hospital, ample to care for the needs of the "clean-up squad", had been estblished in the attending surgeon's former office at the corner of Kaiser Wilhelm Ring and Casino Strasse.

On the very next day-June 4th-orders were received from Washington providing for retention on the Rhine of a force of 1,200 officers and men. It then became necessary to provide better hospital facilities than the attending surgeon's old office afforded. After careful search it was decided to demand a part of the old Station Hospital back from the French. Accordingly, the Surgeon, on June 21st, met General Lasnet, Colonel du Rosselle and members of the French Mission at the Station Hospital, and made plans for the division of the compound, the French and American portions to be separated by a high fence. The arrangement thus made was followed up to the end, the relations between the French hospital authorities and those on the American side continuing to be agreeable and friendly. The hospital moved from the attending surgeon's old office to the compound in Moselweisserweg on June 26th.

Substitutes for Dental and Veterinary Officers.

Upon the departure of the last dental surgeon and last veterinary officer on November 30th, arrangements were made with a German practitioner for the treatment of such emergency dental cases as might be necessary. Major Daniel LeMay, a retired officer of the Veterinary Corps, who was living at Coblenz, took over the veterinary work and devoted his whole time to it without compensation.

Transfer of Attending Surgeon's Office.

After the closing of the attending surgeon's old office at Kaiser Wilhelm Ring and Casino Strasse, all office patients were required to go to the Station Hospital to consult the Attending Surgeon. The tramway service, however, was curtailed to such an extent that this arrangement was unsatisfactory to many. An attending surgeon's office was therefore established in the Headquarters building on November 13th, with office hours from 9 to 11 every morning.

Professional work.

The professional work accomplished during 1922 was of a very high order. Surgery, internal medicine and genito-urinary surgery were all represented by officers of excellent professional knowledge and training. Operations that are seldom attempted in the United States except in large medical centres were successfully performed at the Station Hospital.

Every effort was made to impress on all concerned that the primary function of all Medical Department agencies in time of peace is the care of sick and injured, and that the interests of patients should have first consideration under all circumstances. As a result, serious complaints on the part of patients were almost unknown.

As the Station Hospital of Coblenz was the only institution of its kind in Europe, patients were admitted to it from many different sources, including the American Relief Administration in Russia and the American Embassies at Paris, Berlin, Berne and other capitals. In addition, many Americans traveling in Europe, including members of Congress and other public officials, were cared for. French, British and Belgian officers and soldiers and members of their families were also admitted. The care and attention given the latter class of patients was no small factor in maintaining a spirit of friendliness and good feeling between Americans and Allies in the occupied area.

Sanitation of Command.

Water Supply.

The only new sanitary problem with which the Army medical officers were confronted during the year had to do with the Ehrenbreitstein water supply. The water as delivered at the fort came from a number of groups of shallow wells in the hills behind Arenberg, the flow from which is collected be gravity into a common reservoir. The water at the fort showed continuous pollution, which was finally traced to two wells. The flow from these was excluded from the common supply and no further trouble was experienced.

Kitchens and Mess Halls.

These were almost uniformly excellent from the sanitary point of view.

Food Supplies.

The food supplies furnished by the Quartermaster Department were satisfactory. Issues of meat were regularly inspected by an officer of the Veterinary Corps. When the last reduction of the A.F.G. was made in June, 1922, the shortage of personnel rendered it impracticable to continue the large Army bakery at Coblenz-Luetzel. Arrangements were accordingly made by the Quartermaster with a German bakery by which one pound of flour was furnished for one pound of bread. The bread from this bakery was satisfactory, but the conditions under which it was baked were not so sanitary as those of the old Army bakery.

The Quartermaster Farm continued throughout 1922 to furnish a good milk supply and many products such as fresh vegetables, pork, turkeys and other fowl for the personnel of the A.F.G.

Disposal of wastes.

No change from 1st year.[1]

Laundry.

The Quartermaster laundry at Coblenz-Luetzel was closed in June. The Station Hospital continued to operate a laundry for washing its own linen.

Health of Command.

The general health of the command was excellent. The average non-effective rate per thousand from all causes for the year 1922 was 32.2, as compared with 36. in 1921 and 44. in 1920. Practically no recruits or additions to the garrison were received during the year. As the command therefore consisted entirely of seasoned troops, there were no epidemics. With an average strength of 2,926, there occurred throughout the year the following infectious diseases:

Measles, 9; influenza, 47; pneumonia, 47; diphtheria, 4; scarlet fever, 1; tuberculosis, 4; Vincent's agina, 14; scabies, 74; impetigo contagiosa, 13; gonorrhea, (new), 160; (old), 87; syphilis, (new), 105; (old), 45; chanoroid, 53.

---

[1] See A.R. in O.G., 1920-21, vol II, p. 393.

Venereal Rate.

The Venereal rate continued high. Such was to be expected for the following reasons:

(a) Troops on foreign station have little opportunity of meeting and associating with more respectable women of population;

(b) The pay of soldiers was extremely high when converted into marks;

(c) The poverty of the working-girl in Germany made her susceptible to the very great temptation offered by the fine raiment, automobile drives and suppers enjoyed by those who accepted the soldiers' advances.

All known agencies were utilised to reduce the venereal rate - lectures, prophylaxix, lavish entertainment by the Y.M.C.A., athletics, etc. These undoubtedly did good but were not sufficient to counteract the adverse conditions referred to above.

The average non-effective rate per thousand from venereal disease was 13.87 for 1922, as compared with 16.46 for 1921.

Medical Supplies.

Medical supplies to the value of $363,154.31 were delivered to the agents of the American Relief Administration for use in Russia during the year. Numerous small lots were didposed of by sale to other agencies or through salvage.

After the reduction of the forces in June, the

Medical Supply Depot was moved to a large warehouse in the former Engineer Depot in the compound adjoining the Station Hospital.

It was necessary to make very few requisitions on the United States depots for supplies for these forces.

Civil Affairs.

The medical administration, Civil Affairs, is fully dealt with in sub-chapter C of Chapter 3.

Dental Service.

There is nothing of importance to be noted under this head except, as already mentioned, that the command was without a dental surgeon from November 30th on.

Veterinary Service.

The veterinary service functioned very satisfactorily up to November 30th, when the last veterinary officer on duty with the A.F.G. was withdrawn without replacement. The Commanding General always showed great personal interest in the care given public animals and in their hygiene and surroundings. Stables were kept in excellent condition and feeding, grooming and exercise received the best possible attention. The health of the animals was good in consequence. No epidemics prevailed either in organisation stables or at the Quartermaster farm, where dairy cattle, hogs and poultry were maintained.

The veterinary service, in addition, did valuable work during the year in the inspection of meat, dairy products and forage.

Officers.

The following officers served with the Medical Department, A.F.G., during the periods indicated.

Medical Corps.

Keefer, F.R.	Colonel	Jan.1,1922 to June 20,1922.
Miller, R.B.	Lt.Col.	Jan.1,1922 to Jan.27,1923.
Reasoner, M.A.	Major	Jan.1,1922 to Jan.27,1923.
McKinney, G.L.	"	Jan.1,1922 to June 30,1922.
McDiarmid, N.L.	"	Jan.1,1922 to May 27, 1922.
Davis, A.O.	"	Jan.1,1922 to May 1,1922.
Michie, H.C.	"	Nov.6,1922 to Jan.27,1923.
Flynn, T.J.	"	Jan.1,1922 to May 27,1922.
Walson, C.M.	"	Jan.1,1922 to May 27,1922.
Grant, D.N.W.	"	Jan.1,1922 to April 8,1922.
Dear, W.R.	"	Jan.1,1922 to Jan.17,1922.
Jones, G.I.	"	Jan.1,1922 to Jan.17,1922.
Neff, E.B.	"	Jan.1,1922 to Jan.27,1923.
Francis, F.D.	"	Mar.7,1922 to Jan.27,1923.
Miltenberger, V.E.	"	Janl,1922 to May 19,1922.
Bailey, C.O.	"	Jan.1,1922 to June 30,1922.
McClintic, B.S.	"	Janl,1922 to Feb.23,1922.
Fitts, F.M.	"	Jan.1,1922 to May 19,1922.
Lowry, R.H.	"	Jan.1,1922 to April 8,1922.
Wagner, P.S.	Captain	Mar.22,1922 to April 23,1922.
Brown, H.O.	"	Jan.1,1922 to May 19,1922.
McKeever, J.W.	"	Jan.1,1922 to Mar.19,1922.
Rogers, H.D.	"	Jan.1,1922 to May 16,1922.
Smith, W.A.	"	Jan.1,1922 to April 8,1922.
Tyng, F.C.	"	Jan.1,1922 to May 23,1922.
Price, H.H.	"	Jan.1,1922 to May 19,1922.
Jacques, W.A.	"	Jan.1,1922 to April 8,1922.
Weatherby, F.E.	"	Jan.1,1922 to April 15,1922.
Munly, M.C.	"	Jan.1,1922 to April 1,1922.
Campbell, W.R.	"	Jan.1,1922 to April 8, 1922.
Turner, W.K.	"	Jan.1,1922 to March 8,1922.
Carroll, W.J.	"	Jan.1,1922 to April 8,1922.
Snell, C.F.	"	Jan.1,1922 to May 27,1922.
Calladine, T.M.	"	Jan.1,1922 to May 16,1922.
Tolson, W.F.	"	June 20,1922 to Jan.24,1923.
McNeill, G.P.	"	June 20,1922 to Jan.24,1923.
Gill, W.D.	"	Jan.1,1922 to May 23,1922.

Dental Corps.

Scott, H.O.	Lt.Col.	Jan.1,1922 to May 23,1922.

Gray, D.S.	Captain	Jan.1,1922 to May 16,1922.
Holmes, A.G.	"	Jan.1,1922 to April 8,1922.
Thompson, R.F.	"	Jan.1,1922 to April 23,1922.
Denholm, C.J.	"	Jan.1,1922 to May 19,1922.
Parsons, W.B.	"	Jan.1,1922 to Nov.30,1922.

Veterinary Corps.

Foster, R.J.	Major	Jan.1,1922 to May 27,1922.
Austin, F.H.	1st Lt.	Jan.1,1922 to Nov.30,1922.
Griffin, W.C.	"	Jan.1,1922 to May 16,1922.
Huse, G.B.	"	Jan.1,1922 to April 23,1922.

Medical Administrative Corps.

Cleave, J.W.	1st Lt.	Jan.1,1922 to June 20,1922.
Linden, R.R.	2nd Lt.	Jan.1,1922 to May 27,1922.
Ekman, G.A.	"	Jan.1,1922 to April 8,1922.

SUB-CHAPTER K.

POSTAL EXPRESS SERVICE.

Organization.

Many changes occurred in the Postal Express Service during the year 1922.[1] Chief among these were discharge of practically all the civilian personnel, demobilization of the Postal Detachment, reduction of the enlisted personnel of the Express Service and change in routing of mail.

With the impending departure of the Second Brigade in the summer of 1922, a large reduction was made in the effectives of the Postal Express Service. The Civil Service employees who had been taken on the Post Office Department payrolls in March, 1921, were dropped therefrom on February 1st, 1922. No War Department funds being available to retain any of the Civil Service men, they were all returned to the United States in March. The Postal Express Service fortunately had enlisted personnel of sufficient experience to perform the work for the reduced force of the A.F.G. The Postal Department retained the postal agent to supervise the sale of money orders and stamps and to advise on postal matters.

In compliance with General Order #34, Headquarters, A.F.G., dated March 8th, 1922, the Postal Detachment was demobilized and the members were transferred in their grades and ratings to Headquarters Troop, A.F.G., from which organization they were placed on detached service with the P.E.S.

---

[1]Full details of organization and early history of P.E.S. contained in Chapter 27, A.R. in O.G., 1920-21.

The enlisted personnel was reduced from fourteen to four men in June, 1922. On June 1st, 1922, the Post Office was moved from the four rooms occupied in the Oberpostdirektionsgebauede and all departments were consolidated in one room in Headquarters Building. This, following the reduction of the A.F.G. and the decrease in the Post Office personnel, was of distinct advantage in the supervision and execution of work. The force remaining on the day of transfer of the American area to French control consisted of the Postal Officer, three enlisted men and the postal agent, Post Office Department.

Courier Service.

The ase Post Office at A ntwerp was abandoned on the closing of that port in June, 1922, and the courier service, which had been maintained daily between A.P.O. 927 and A.P.O. 944, was discontinued. The function of A.P.O. 702, Paris, continued without interruption throughout the year. Pouches were exchanged daily between Headquarters A.F.G., and Paris by the French diplomatic courier service, and between these Headquarters and the British A rmy Post Office at Cologne by British courier.

Change in Routing of Mail.

An important change in the mail service of the A.F.G. took place after the Port of Antwerp was closed. Arrangements were completed to send and receive mail on boats at both Hamburg and Bremen, sealed sacks containing U.S. mails from and to the ports being transported by the German Postal authorities under the Universal Postal Union regulations. This service proved satisfactory, in that mails were received more frequently than they had been by way of Antwerp. Whereas the average number of mails recieved prior to the change was four dispatches per month, the arrangement effected by the change furnished an average of eight per month. Mail was dispatched to the United States on an average of twice a week.

A.F.G. Customs Office.

Acting under authority contained in a letter from the Secretary of the Treasury, a customs office was placed in operation at A.P.O. 927 on November 15th and continued until December 30th for the collection of customs on Christmas gifts sent to the United States by members of the A.F.G. The business transacted is indicated by the following figures:

Number of parcels sent duty prepaid	741
Number of parcels valued at less than one dollar	937
Value of parcels duty prepaid	$2,904.37
Value of parcels under dollar	$ 586.84
Total amount of customs duty and internal revenue monies collected. . .	$1,454.63

In summing up the year's operation of the Postal Express Service, it may be said that the service was remarkably free from complaints of loss or damage to mail matter dispatched from or received at A.P.O. 927, and that the reception and dispatch of mail through Hamburg and Bremen provided the A.F.G. with a better service than that enjoyed theretofore.

Officers.

The following officers were on duty as Postal Officers during 1922-23:

Capt. R.P. Bell, Infantry, January 1, 1922-June 17,1922
1st Lt. J.M.Glasgow, Cav., June 18, 1922 - July, 25,1922
1st Lt. M.A.Devine, Cav., July 26, 1922 -Feb, 14, 1923.

DATA ON P.E.S. TRANSACTIONS, JANUARY 1.

1922 - December 31, 1922.

Receipt and Dispatch of Mail.

Number of dispatches received	85
Number of dispatches made	104
Number of pouches first class received	242
Number of sacks second class and P.P. received	1201
Number of pouches first class dispatched	474
Number of sacks second class and P.P. dispatched	1001

Money Orders, Stamps, Insurance and Registry Business.

Number of money orders issued	3201
Number of money orders paid	654
Value of money orders issued	$121,427.64
Value of money orders paid	21,935.37
Total stamp sale	4,500.03
Number of parcels insured	1635
Number of registers received	6357
Number of articles registered and forwarded	1121

SUB-CHAPTER L.

MISCELLANEOUS

Employment of Civilians.

As the reduction of the forces caused a scarcity of enlisted personnel qualified to perform certain work necessary to the proper functioning of the forces, it became necessary to continue the employment of civilians in various capacities. Civilians thus employed included American, Allied and German. They were assigned to all the classes of work performed by clerks, stenographers, skilled mechanics, accountants, chauffeurs, etc. A great majority of the civilians employed, however, were common laborers used by the various corps and departments and by units within certain combatant organizations.

American and Allied civilians in numbers proportionate to the requirements of the reduced command were kept in the employ of the Army throughout 1922. Most of these were employed in a clerical capacity. In the employment of civilians other than German, an effort was made to give preference to American nationals. However, during the early part of the year, complaints came to the attention of the Commanding General that this policy was not being followed. The following order, published in the "Amaroc" and given publicity in other ways within the area, shows the attitude taken by Headquarters, A.F.G., in this matter:

"Headquarters,
American Forces in Germany

February 4, 1922.

Citizens of the United States who consider that they have been discriminated against as regards employment by any branch of the American Forces in Germany, in favor of citizens of other nations, are invited to submit to these Headquarters any complaints that they may have, for consideration and investigation, in connection with an inquiry now being made as regards this matter.

By Command of Major General ALLEN:

C.M. Thiele,
Acting Adjutant"

German civilian labor continued to be employed throughout the year. With regard to the method of employment and the supervision of labor, such as that employed by the Quartermaster Corps, Ordnance and Medical Departments, etc., there was no change. To keep abreast of the ever-changing value of the mark and the increasing cost of living in the American zone, wage scales were frequently adjusted by the Fourth Section, in conjunction with the supervising German agencies. In August, the German government assumed the expenses incident to the employment of civilian labor and thereafter paid wages on a dollar basis. The civilian employees working for the Army and paid by the German government after August, varied from 955 on January 1st, 1922, to 867 on July 1st, and 963 on January 1st, 1923. These numbers included the services of American, Allied and German personnel in all classes of work. German labor employed by companies, batteries, detachments, etc., as janitors, kitchen workers, etc., was paid with funds contributed by the officers and command of the units concerned.

Service Clubs.

The Staff N.C.O. Club, for grades 1-3 inclusive, and the Sergeants' Club, for grades 1 - 4, inclusive, both of which were opened in 1920, found it impossible to continue their respective activities with the reduced personnel and ceased to exist in May, 1922. The rules and policies governing the operation of these clubs are set forth in detail in the 1920-21 volume. They were not changed during the period covered by this report.

The Field Clerks' Club was maintained till the withdrawal of the forces. This little club, opened in 1920, was an important factor in the life and welfare of those members of the A.F.G. who, on account of their status, did not derive the benefits of either the clubs for officers, or those for enlisted men.

Civilian Clubs.

Both the American Legion and the Veterans of Foreign Wars maintained club rooms at Coblenz which were open to members of the A.F.G. under certain conditions. These clubs remained open till the departure of the forces.

Army Library.

The library, maintained by the Y.M.C.A., continued to play an important part in the life and welfare of the Allied colony of Coblenz. Pursuant to the policy of releasing to the Germans all buildings possible, the library was moved from its old quarters into rooms in the Festhalle, where, after July, it continued to operate until January, 1923. In addition to the central library, the Y.M.C.A. maintained smaller libraries at their various "huts" and, in this way, were able to reach a large number of soldiers who, for various reasons, did not patronize the main library.

The records of the library show that during the last six months of the year, a total of 15,397 books were circulated, of which approximately 3,600 were non-fiction, such as philosophy, psychology, religion, sociology, mathematics, art, music, sport, travel, history, biography, poetry, drama, essays, etc. Each month brought new readers to the library. Its patrons included not only officers, men and families of the A.F.G., but British, French and Belgians, both military and civilian.

In addition to the large number of books at the library, current magazines of all descriptions were to be found in its reading rooms. During the last half of the year, 14,218 military and 14,152 civilians made use of the rooms.

American Visitors.

The influx of American civilians and of retired Army officers into the Coblenz bridgehead continued as in former years. The desire to see the American flag waving over Ehrenbreitstein, no doubt, represented the sentimental aspect of this sojourn; on the other hand, the many advantages afforded in the American area, where the military authorities did everything possible to make the travelers comfortable, probably furnished the chief attraction. Newspaper correspondents and free-lance feature agents were also much in evidence. All received most courteous treatment; even the occasional critic who took it upon himself to investigate the A.F.G. on his own account was given due consideration.

Of the retired Army officers, who visited Coblenz off and on during the year, the most distinguished were: Major General Peyton C. March, Major General John L. Chamberlain, and Brigadier General Charles G. Treat.

Major General Frank McIntyre, Chief of the Bureau of Insular Affairs, while traveling in Europe in 1922, also visited the area.

Made in the USA
Coppell, TX
05 November 2026

11315630R00273